ZnO Nanostructures for Tissue Regeneration, Drug-Delivery and Theranostics Applications

ZnO Nanostructures for Tissue Regeneration, Drug-Delivery and Theranostics Applications

Editors

Valentina Cauda
Marco Laurenti

MDPI • Basel • Beijing • Wuhan • Barcelona • Belgrade • Manchester • Tokyo • Cluj • Tianjin

Editors
Valentina Cauda
Department of Applied Science
& Technology, Politecnico di
Torino, Corso Duca degli
Abruzzi 24
Italy

Marco Laurenti
Applied Science and Technology
Department, Politecnico di
Torino, Corso Duca degli
Abruzzi 24
Italy

Editorial Office
MDPI
St. Alban-Anlage 66
4052 Basel, Switzerland

This is a reprint of articles from the Special Issue published online in the open access journal *Nanomaterials* (ISSN 2079-4991) (available at: https://www.mdpi.com/journal/nanomaterials/special_issues/ZnO_tissue).

For citation purposes, cite each article independently as indicated on the article page online and as indicated below:

LastName, A.A.; LastName, B.B.; LastName, C.C. Article Title. *Journal Name* **Year**, *Volume Number*, Page Range.

ISBN 978-3-0365-0656-2 (Hbk)
ISBN 978-3-0365-0657-9 (PDF)

Cover image courtesy of Valentina Cauda.

© 2021 by the authors. Articles in this book are Open Access and distributed under the Creative Commons Attribution (CC BY) license, which allows users to download, copy and build upon published articles, as long as the author and publisher are properly credited, which ensures maximum dissemination and a wider impact of our publications.
The book as a whole is distributed by MDPI under the terms and conditions of the Creative Commons license CC BY-NC-ND.

Contents

About the Editors vii

Valentina Cauda and Marco Laurenti
Editorial for Special Issue: ZnO Nanostructures for Tissue Regeneration, Drug-Delivery and Theranostics Applications
Reprinted from: *Nanomaterials* **2021**, *11*, 296, doi:10.3390/nano11020296 1

Marina Martínez-Carmona, Yurii Gun'ko and María Vallet-Regí
ZnO Nanostructures for Drug Delivery and Theranostic Applications
Reprinted from: *Nanomaterials* **2018**, *8*, 268, doi:10.3390/nano8040268 5

Nadia Garino, Tania Limongi, Bianca Dumontel, Marta Canta, Luisa Racca, Marco Laurenti, Micaela Castellino, Alberto Casu, Andrea Falqui and Valentina Cauda
A Microwave-Assisted Synthesis of Zinc Oxide Nanocrystals Finely Tuned for Biological Applications
Reprinted from: *Nanomaterials* **2019**, *9*, 212, doi:10.3390/nano9020212 33

Rebeca Pérez, Sandra Sanchez-Salcedo, Daniel Lozano, Clara Heras, Pedro Esbrit, María Vallet-Regí and Antonio J. Salinas
Osteogenic Effect of ZnO-Mesoporous Glasses Loaded with Osteostatin
Reprinted from: *Nanomaterials* **2018**, *8*, 592, doi:10.3390/nano8080592 51

Mariusz Cierech, Izabela Osica, Adam Kolenda, Jacek Wojnarowicz, Dariusz Szmigiel, Witold Łojkowski, Krzysztof Kurzydłowski, Katsuhiko Ariga and Elżbieta Mierzwińska-Nastalska
Mechanical and Physicochemical Properties of Newly Formed ZnO-PMMA Nanocomposites for Denture Bases
Reprinted from: *Nanomaterials* **2018**, *8*, 305, doi:10.3390/nano8050305 69

Federica Leone, Roberta Cataldo, Sara S. Y. Mohamed, Luigi Manna, Mauro Banchero, Silvia Ronchetti, Narcisa Mandras, Vivian Tullio, Roberta Cavalli and Barbara Onida
Nanostructured ZnO as Multifunctional Carrier for a Green Antibacterial Drug Delivery System—A Feasibility Study
Reprinted from: *Nanomaterials* **2019**, *9*, 407, doi:10.3390/nano9030407 83

About the Editors

Valentina Cauda is Associate Professor at the Department of Applied Science and Technology (DISAT), Politecnico di Torino, head of the TrojaNanoHorse lab (in brief TNHLab) and co-founder of the Interdepartmental laboratory PolitoBIOMed Lab. Thanks to her ERC Starting Grant project (TrojaNanoHorse, GA 678151), established in March 2016, she now leads a multidisciplinary research group of 18 people, including chemists, biologists, physics, engineers, and nanotechnologists. Her main research topic is theranostic nanomaterials: the research team develops metal oxide nanomaterials from wet synthesis, chemical functionalization, and physical–chemical characterization up to their coating by lipidic bilayer from both artificial and natural origins, aimed toward drug delivery, tumor cell targeting, and bio-imaging. Metal oxide nanomaterials, like zinc oxide, mesoporous silica, titania, and metal (gold, silver) nanostructures as well as liposomes and cell-derived extracellular vesicles, are investigated. Valentina Cauda graduated in Chemical Engineering in 2004 from Politecnico di Torino and then received her PhD in Materials Science and Technology in 2008. After a short period at the University of Madrid, she worked as a Postdoc at the University of Munich, Germany, on nanoparticles for drug delivery and tumor cell targeting. From 2010 to 2015, she was Senior Postdoc at the Istituto Italiano di Tecnologia in Torino, followed by a move to Politecnico di Torino, where she was appointed Associate Professor. In recognition of her research, she received the prize for young researchers at the Chemistry Department of the University of Munich in 2010, the Giovedì Scienza award in 2013, the Zonta Prize for Chemistry in 2015, and the USERN Prize for Biological Sciences in 2017. She has 113 scientific publications and a h-index of 36 (updated on January 2021). She holds 4 international patents on the use of metal oxide nanoparticles in nanomedicine. Prof. Cauda is principal investigator of several industrial, national, and international projects for which she has collectively raised over 5 M€ in funding. The most relevant are the recently granted ERC Proof-of-Concept XtraUS N. 957563, the FET Open RIA MIMIC-KEY, the Marie Skłodowska-Curie Action MINT N. 842964 (where she acts as supervisor of an incoming postdoc from abroad), and the ERC Starting Grant TrojaNanoHorse. More details are available at https://areeweb.polito.it/TNHlab/.

Marco Laurenti received his MSc degree in Physical Engineering in 2011 from the Politecnico di Torino. His thesis work focused on the development of biocompatible thin films (a-Si:H, ZnO) for promoting osteoblasts cell adhesion. In February 2015, he received his PhD in Physics from the Politecnico di Torino in collaboration with the Istituto Italiano di Tecnologia, Center for Space Human Robotics. The subject of his PhD thesis was the sputtering deposition and characterization of pristine and doped ZnO piezoelectric thin films for sensing and energy-harvesting applications. He is currently working as a research assistant at the Politecnico di Torino. His activities and research interests include sputtering deposition of metal oxide thin films (biocompatible/bioactive porous films, electrochromic materials, thin films for memristive devices) and CVD growth of single-layer graphene as a nanoporous membrane for seawater desalination and wastewater treatment.

Editorial

Editorial for Special Issue: ZnO Nanostructures for Tissue Regeneration, Drug-Delivery and Theranostics Applications

Valentina Cauda * and Marco Laurenti *

Department of Applied Science & Technology, Politecnico di Torino, Corso Duca degli Abruzzi 24, 10129 Torino, Italy
* Correspondence: valentina.cauda@polito.it (V.C.); marco.laurenti@polito.it (M.L.)

Received: 13 January 2021; Accepted: 21 January 2021; Published: 24 January 2021

In recent years, zinc oxide (ZnO)-based nanomaterials have attracted a great deal of interest thanks to their outstanding and multifunctional properties. Actually, ZnO can be synthesized in a broad variety of nano-sized morphologies (such as nanowires, nanorods, nanoparticles, and nanoflowers), shows easy preparation routes and facile surface chemical functionalization. Most importantly, ZnO has many intriguing properties, being a semiconductor, piezoelectric, pyroelectric and photoexcitable material, with low chemical stability in acidic environments and interesting antimicrobial and anticancer properties. These aspects fostered a deep investigation of ZnO nanomaterials to design and fabricate smart biocompatible nanotools, which have been successfully applied to a wide plethora of applications in the biomedical field. In such cases, ZnO nanostructures, alone or combined into hybrid or composite systems, represent a powerful tool for the fabrication of new scaffolds for tissue regeneration with improved antimicrobial properties, as well as for drug delivery applications. Moreover, the promising optical and biocompatible properties of ZnO have been successfully combined together, resulting in the co-presence of imaging and therapeutic actions, useful for theranostics applications towards cancer therapy.

This Special Issue of *Nanomaterials* is therefore dedicated to the most recent advances in the use of ZnO nanostructures for designing novel smart nanomaterials dedicated to biomedical systems, tissue engineering, drug delivery and theranostics devices. It ranges from the synthesis and characterization of the starting nanomaterials, to their final in vitro applications.

To have a proper overview in the specific field of ZnO nanostructures for drug delivery and theranostics applications, the Review from Prof. Maria Vallet-Regì and coworkers [1] is very relevant. Here, the authors analyze recent strategies in proposing ZnO as semiconductor quantum dots (QDs) not only for bio-imaging purposes but also as multifunctional tools, i.e., for drug delivery and theranostic imaging against different diseases. In particular, the application of ZnO for antibacterial or anti-inflammatory treatments, against diabetes and cancer, as well as in wound healing are proposed with various in vitro and in vivo examples from the literature.

In the paper of Dr. Nadia Garino et al. [2], a special focus is given to the synthetic protocols applied to produce ZnO nanocrystals and their surface decoration by aminopropyl groups facing colloidal dispersion and stability over time when used towards cancer cells. Actually, the paper shows a novel microwave-assisted sol–gel synthetic route, pointing out how important it is to control all the nanomaterial properties when dealing with biological entities, i.e., living cells, for the achievement of reproducible and reliable results.

A similar relationship between the nanomaterials' properties, the synthetic route and the interaction with the biological world (here microorganisms) is reported in the work of Prof.s Roberta Cavalli, Barbara

Onida and coworkers [3]. Therein, wet organic-solvent-free processes were used to produce two ZnO nanostructures with different morphologies, yet providing different surface areas, crystal sizes, and thus dissolution rates into zinc cations. The antimicrobial effects of these ZnO nanostructures were then measured on various bacterial strains and the successful loading of the anti-inflammatory drug ibuprofen was successfully proposed for the first time using a supercritical CO_2-mediated impregnation process. This paper demonstrates the potential use of ZnO nanomaterials as a multifunctional antimicrobial drug nanocarrier.

Concerning bone tissue engineering applications, the work of Prof. Maria Vallet-Regì and Antonio Salinas [4] shows that ZnO can be efficiently used in Mesoporous Bioactive Glasses (MBGs) as carriers for the peptide osteostatin. Interestingly, the zinc cations release from the MBG, combined with the osteogenic properties of osteostatin, provided a valuable tissue engineering device, as proved by in vitro tests with pre-osteoblasts.

Another representative in the field of tissue engineering is the work of M. Cierech et al. [5]; in this study, ZnO nanoparticles incorporated into a polymeric matrix were successfully designed to simultaneously show anti-bacterial effects and retention of both mechanical and hydrophilic properties useful for preparing a denture base.

As a concluding remark, with this Special Issue we hope we have contributed to highlight the role of zinc oxide nanomaterials in cancer cell theranostics, drug delivery and tissue engineering, providing insights from their synthesis, surface functionalization and characterization to their smart behaviors with customizable properties for advanced and personalized biomedical applications.

Author Contributions: Conceptualization, V.C.; writing—original draft preparation, V.C.; writing—review and editing, V.C. and M.L.; supervision, V.C. and M.L. Both authors have read and agreed to the published version of the manuscript.

Funding: This research received no external funding.

Conflicts of Interest: The authors declare no conflict of interest.

References

1. Martinez-Carmona, M.; Gun'ko, Y.; Vallet-Regí, M. ZnO Nanostructures for Drug Delivery and Theranostic Applications. *Nanomaterials* **2018**, *8*, 268. [CrossRef] [PubMed]
2. Garino, N.; Limongi, T.; Dumontel, B.; Canta, M.; Racca, L.; Laurenti, M.; Castellino, M.; Casu, A.; Falqui, A.; Cauda, V. A Microwave-Assisted Synthesis of Zinc Oxide Nanocrystals Finely Tuned for Biological Applications. *Nanomaterials* **2019**, *9*, 212. [CrossRef] [PubMed]
3. Leone, F.; Cataldo, R.; Mohamed, S.S.Y.; Manna, L.; Banchero, M.; Ronchetti, S.; Mandras, N.; Tullio, V.; Cavalli, R.; Onida, B. Nanostructured ZnO as Multifunctional Carrier for a Green Antibacterial Drug Delivery System—A Feasibility Study. *Nanomaterials* **2019**, *9*, 407. [CrossRef] [PubMed]
4. Pérez, R.; Sanchez-Salcedo, S.; Lozano, D.; Heras, C.; Esbrit, P.; Vallet-Regí, M.; Salinas, A.J. Osteogenic Effect of ZnO-Mesoporous Glasses Loaded with Osteostatin. *Nanomaterials* **2018**, *8*, 592. [CrossRef] [PubMed]
5. Cierech, M.; Osica, I.; Kolenda, A.; Wojnarowicz, J.; Szmigiel, D.; Łojkowski, W.; Kurzydłowski, K.; Ariga, K.; Mierzwińska-Nastalska, E. Mechanical and Physicochemical Properties of Newly Formed ZnO-PMMA Nanocomposites for Denture Bases. *Nanomaterials* **2018**, *8*, 305. [CrossRef] [PubMed]

Publisher's Note: MDPI stays neutral with regard to jurisdictional claims in published maps and institutional affiliations.

 © 2021 by the authors. Licensee MDPI, Basel, Switzerland. This article is an open access article distributed under the terms and conditions of the Creative Commons Attribution (CC BY) license (http://creativecommons.org/licenses/by/4.0/).

Review

ZnO Nanostructures for Drug Delivery and Theranostic Applications

Marina Martínez-Carmona [1], Yurii Gun'ko [1] and María Vallet-Regí [2,3,*

1. School of Chemistry and CRANN, Trinity College, The University of Dublin, Dublin 2, Ireland; martim10@tcd.ie (M.M.-C.); igounko@tcd.ie (Y.G.)
2. Department Chemistry in Pharmaceutical Sciences, School of Pharmacy, Universidad Complutense de Madrid, Instituto de Investigación Sanitaria Hospital 12 de Octubre i+12, 28040 Madrid, Spain
3. Centro de Investigación Biomédica en Red de Bioingeniería, Biomateriales y Nanomedicina (CIBER-BBN), Avenida Monforte de Lemos, 3-5, 28029 Madrid, Spain
* Correspondence: vallet@ucm.es; Tel.: +34-913941861; Fax: +34-913941786

Received: 13 March 2018; Accepted: 18 April 2018; Published: 23 April 2018

Abstract: In the last two decades, zinc oxide (ZnO) semiconductor Quantum dots (QDs) have been shown to have fantastic luminescent properties, which together with their low-cost, low-toxicity and biocompatibility have turned these nanomaterials into one of the main candidates for bio-imaging. The discovery of other desirable traits such as their ability to produce destructive reactive oxygen species (ROS), high catalytic efficiency, strong adsorption capability and high isoelectric point, also make them promising nanomaterials for therapeutic and diagnostic functions. Herein, we review the recent progress on the use of ZnO based nanoplatforms in drug delivery and theranostic in several diseases such as bacterial infection and cancer.

Keywords: ZnO nanoparticles; Quantum dots; theranostic; drug delivery; anti-tumour; diabetes treatment; anti-inflammation; antibacterial; antifungal; wound healing

1. Introduction

For many years, the use of organic dye molecules has allowed us to detect and monitor various kinds of substances, including drugs, amino acids, nucleotides or materials, both in and outside of cells. They have also been used to study the process of life chemistry (enzymatic synthesis, immune response, etc.) or to identify some diseases. However, its use in bio-imaging has been drastically reduced since the appearance of the quantum dots (QDs) [1]. In general, QDs are more stable to photochemical degradation, have wide excitation wavelength ranges and narrow and symmetric emission spectra and can exhibit different colours depending on the size of the particle [2] (the so-called quantum size effect) [3].

Among the typical QDs, i.e., CdSe, CdTe, CuO, TiO$_2$, etc., ZnO are without any doubt one of the best choices since they are excellent semiconductors, with luminescent properties [4,5] almost as good as those of the Cd QDs ones but presenting the advantage of being biodegradable and nontoxic [6]. In fact, although its effect at the nanometre level has not yet been established, ZnO in bulk has already been considered as safe and approved by the US Food and Drug Administration [7].

ZnO is an n-type semiconductor with an outsized exciton-binding energy (60 meV), a wide band gap of 3.37 eV at room temperature, a Bohr exciton radius of ~2.34 nm and a high dielectric constant. Irradiatation of ZnO with UV light favours the promotion of an electron (e$^-$) to the conduction band and therefore producing a hole (h$^+$) in the valence band, namely the electron/hole pair. Apart from this typical UV range excitonic emission, the photoluminescence spectrum of ZnO nanocrystals also displays a broad visible emission, more suitable for biological imaging. This extended emission has been ascribed to point defects such as O and Zn vacancies or interstitials and related to surface oxygen-containing moieties, such as OH groups [8].

Moreover, the luminescence of ZnO nanocrystals can be improved or modulated by doping the structure with other ions [9,10].

In addition, ZnO QDs also present properties such as the ability to produce ROS, a strong adsorption capability and an easily tuneable surface that play a crucial role in their use for biomedical applications.

When ZnO crystals are under UV irradiation in aqueous suspension, these electron/hole pairs will produce several photochemical reactions generating ROS, making them good candidates for photodynamic therapy [11]. Usually, when ZnO QDs are excited, the valence band holes present on the surface, abstract electrons from water and/or hydroxyl ions, giving place to hydroxyl radicals (OH$^\bullet$). At the same time, the superoxide anion O_2^- is produced due to the reduction of oxygen [12]. Apart from the high production of ROS after UV irradiation, ZnO QDs themselves can also generate small amounts of ROS due to the pro-inflammatory response of the cell against nanoparticles (NPs) [13] and to the characteristic surface property of ZnO QDs [14,15]. Normally, UV light is required to produce these electron/hole pairs, however, for ZnO particles whose size is on the nanometre scale, electrons can also reach the conduction band without the help of UV excitation [16], probably because of the presence of crystal defects due to their nano-size. Fortunately, this phenomenon is of little importance outside the cells, where the concentration of ROS is small, but once internalized higher levels of ROS resulting in cell death. Some studies reveal that ROS production is significantly higher in tumour cells than in normal ones after being treated with ZnO QDs [17]. It has been reported that various signalling molecules and ROS are generally more abundant in cells such as tumour cells, due to a rapid metabolic rate, and high degree of growth and multiplication than in normal cells [18].

ZnO QDs have also be produce a variety of different nano-architectures, including nanospheres, nanorods, nanotubes, nanorings, nanobelts, nanoflowers, etc. [19–23].

ZnO QDs have a versatile surface chemistry that can easily be modified to prevent aggregation, improve colloidal stability [24] or to obtain new properties as drug delivery systems (DDS) [25,26].

The use of DDS in nanomedicine has important advantages compared with traditional drugs: (i) increasing solubility of drugs that cannot be taken up by cells and increasing therefore their bioavailability [27,28]; (ii) avoiding the degradation of some drugs that are unstable at physiological or gastrointestinal pH [29,30]; and (iii) reducing the toxicity and side effects of drugs by using target molecules that increase the selectivity of the treatment [31,32].

Considering their ability to produce ROS, capacity to act as drug delivery systems and their luminescence properties, we can talk about theranostic nanoplatforms where ZnO QDs not only perform the role of image agents but also of treatment [33,34].

Herein, in this review, we will summarize the recent progress on the use of ZnO QDs for drug delivery and theranostic imaging in different pathologies (Figure 1). Approaches to the preparation and chemical functionalization of ZnO nanostructures for biological applications are very well documented and this area was a subject of several recent reviews [7,35–39], therefore we are not going to consider these aspects in this manuscript.

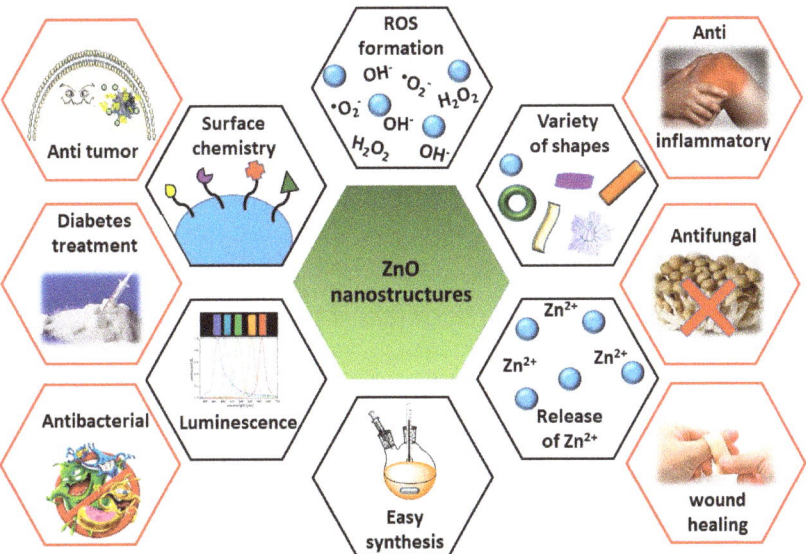

Figure 1. Diagram summarizing the main characteristics of ZnO nanostructures (black hexagons) and their principal applications in biomedicine (red hexagons).

2. ZnO Nanoplatforms for Theranostic in Cancer

In view of the number of publications over the last years, there is no doubt that cancer is the main objective in terms of the use of ZnO based materials for the treatment of diseases. ZnO QDs slowly dissolve in physiological pH [40] producing small changes in extracellular zinc concentrations that cause very little cytotoxicity. However, NPs preferentially internalized in tumour cells as consequence of the enhanced permeability and retention (EPR) effect. Once inside and because of electrostatic interactions, ZnO QDs present certain cytotoxicity by themselves based on a higher intracellular release of dissolved zinc ions due to the acidification of the media, followed by increased ROS induction. This situation results in the loss of protein activity balance mediated by zinc as well as in an oxidative stress environment that finally produce cell death [41].

In addition to the synthetic versatility of these materials, we find that ZnO can act as a core, as a shell around other types of particles or provide an added value to more complex systems. All the systems are summarized and referenced in Table 1.

Table 1. ZnO nanoplatforms for theranostic in Cancer.

Type of Cell/Animal Used [a]	Type of Device [b]	Responsive Phenomena [c]	Drug/Antibiotic [d]	Reference
MCF-7	ZnO QDs	-	Adsorbed DOX	[42]
MCF-7R, MCF-7S	ZnO QDs	pH	Loaded DOX	[43]
MDA-MB-231, HeLa, NCI/ADR-RES, MES-SA/Dx5	ZnO QDs	pH	Adsorbed DOX	[44]
-	ZnO QDs	pH, ultrasounds	Loaded DOX	[45]
HeLa	FA Mg ZnO QDs	pH	Adsorbed DOX	[46]
MCF-7, MDA-MB-231, nude mice	FA Hollow ZnO NPs	pH	Loaded paclitaxel	[47]
MDA-MB-231, HBL-100, mice	FA ZnO Nanosheets	pH, heat	Loaded DOX	[48]
SMMC-7721	ZnO nanorod	UV radiation	-	[12]
HeLa, PC3	Lanthanide–ZnO QDs	UV, X-ray, γ-ray radiation	-	[33]
SMMC-7721	ZnO nanorod	UV radiation	DOX complex	[49]
HNSCC	ZnO QDs	UVA irradiation	Paclitaxel, cisplatin	[50]

Table 1. Cont.

Type of Cell/Animal Used [a]	Type of Device [b]	Responsive Phenomena [c]	Drug/Antibiotic [d]	Reference
MCF-7	MUC1 aptamer S2.2. ZnO QDs	UV radiation	Loaded DOX	[51]
BxPC-3, tumour-bearing nude mice	Gd-Polymer–ZnO QDs	pH	Adsorbed DOX	[34]
HEK 293T, HeLa	FA-SiO$_2$ ZnO NPs	pH	Loaded DOX	[52]
HeLa	Lipid ZnO NCs	pH	-	[53]
Caco-2	TiO$_2$@ZnO–GO and TiO$_2$@ZnO	pH	Loaded Cur	[26]
-	Fe$_3$O$_4$@ZnO@mGd$_2$O$_3$:Eu@P(NIPAm-co-MAA)	Microwave, Magnetic radiation	VP-16	[54]
MCF-7	β-CD-Fe$_3$O$_4$@ZnO: Er^{3+}, Yb^{3+}	Microwave, Magnetic radiation	VP-16	[55]
HeLa	ZnO MSNs	pH	Loaded DOX	[56]
BxPC-3	Mg ZnO MSNs	pH	Loaded CPT, adsorbed Cur	[57]
HeLa, mouse	UCNPs@mSiO$_2$-ZnO	pH	Loaded DOX	[58]
-	ZnO-pSiO$_2$-GSSG NPs	Protease, redox, pH	Loaded amoxicillin	[59]
HepG$_2$	L-pSiO$_2$/Cys/ZnO NPs	Redox, pH	Loaded DOX	[25]
A549	ZnO-MCNs	pH	Loaded MIT	[60]
HeLa	ZnO@-Dextran microgels	pH	Loaded DOX	[61]

[a] MCF-7: Human breast cancer cell; MCF-7S/MCF-7R: Human breast cancer cell sensitive/resistant to doxorubicin; MDA-MB-231: epithelial, human breast cancer cell; HeLa: Human epithelial cells from a fatal cervical carcinoma; NCI/ADR-RES: Ovarian tumour cell; MES-SA/Dx5: Multidrug-resistant human sarcoma cell; HBL-100: Human, Caucasian, breast cancer cell; SMMC-7721: Human hepatocarcinoma cell; PC3: Human prostate cancer cell; HNSCC: Head and neck squamous cell carcinoma; BxPC-3: Human pancreatic cancer cell; HEK 293T: Human embryonic kidney cells; Caco-2: Human epithelial colorectal adenocarcinoma cell; HepG$_2$: Human liver cancer cell; A549: Adenocarcinomic human alveolar basal epithelial cell. [b] ZnO QDs: Zinc oxide quantum dots; FA: Folic acid; QDs: Quantum dots; MUC1: membrane glycoprotein which is highly expressed in most breast cancers; Aptamer S2.2.: (5′-COOH-GCA-GTT-GAT-CCT-TTG-GAT-ACC-CTGGTTTTT-FAM-3′) SiO$_2$: Silica; NCs: Nanocrystals; MABG: TiO$_2$@ZnO–GO: ZnO coated mesoporous titanium oxide QDs containing graphene oxide; Fe$_3$O$_4$@ZnO@mGd$_2$O$_3$:Eu@P(NIPAm-co-MAA): iron oxide QDs coated with ZnO and mesoporous Gd$_2$O$_3$:Eu shells with a polymer poly[(N-isopropylacrylamide)-co-(methacrylic acid)] (P(NIPAm-co-MAA)) to gate the mesoporous; K8(RGD)2 cationic peptide containing 2 RGD sequences; β-CD-Fe$_3$O$_4$@ZnO: Er^{3+}, Yb^{3+}: β-cyclodextrins functionalized iron oxide QDs doped with Er^{3+} and Yb^{3+} coated with ZnO; ZnO MSNs: Mesoporous silica nanoparticles with ZnO QDs as cap of the pores; UCNPs@mSiO$_2$-ZnO: Lanthanide-doped upconverting nanoparticles with a mesoporous silica layer and ZnO QDs as gatekeeper; ZnO-pSiO$_2$-GSSG NPs: ZnO QDs as cups of oxidized glutathione (GSSG) amino-functionalized silica NPs; L-pSiO$_2$/Cys/ZnO NPs: Lemon like silica NPs with cysteine and ZnO QDs cups; MCNs: Mesoporous carbon nanoparticles. [c] UV: Ultra violet. [d] DOX: Doxorubicin; Cur: Curcumin; VP-16: Chemotherapeutic drug etoposide; CPT: Camptothecin; and MIT: Mitoxantrone.

2.1. ZnO Core Nanosystems

Regarding systems that use ZnO as core for the treatment of cancer, in 2016, Vaidya et al. adsorbed doxorubicin (DOX) onto the surface of ZnO QDs (ZD QDs) and studied their anticancer activity in MCF-7 cells compared to that presented by free DOX, ZnO QDs, and a mixture of the latter two. It was observed that the combined addition of ZnO QDs and DOX presented higher antitumour capacity than any of its components separately but lower than the effect of ZD QDs, maybe because of a better targeting and a higher retention of the DOX loaded QDs in the tumour cells [42]. At the same time, Liang et al. performed a similar study with MCF-7R and MCF-7S cells. In this case, they explained the release of the drug due to the degradation of ZnO in response to pH after internalization of the QDs into the endo/lysosomes. They also performed a real-time tracking of the drug release. Although the two components separately exhibited fluorescence, their intensity was quenched after ZD QDs formation. However, after the degradation of ZnO and the consequent release of the DOX, the intensity of the fluorescence increased again [43]. In 2017, Zhu et al. went one step further and proposed ZnO QDs as a multifunctional platform for cancer treatment (Figure 2a). They studied the synergistic anticancer activity due to the ROS generation of ZnO QDs and DOX in several cell lines but also studied their effect in macrophages or in tumour (stem-like) cells. In stem cells, it was observed that ZnO QDs affected the expression of CD44, leading to a marked decrease in migration, accumulation of mutations and cell adhesion, but increasing sensitivity to antitumour treatment. In macrophages, a polarization towards the phenotype M1 was observed, increasing the antitumour effectiveness and immune response of DOX [44].

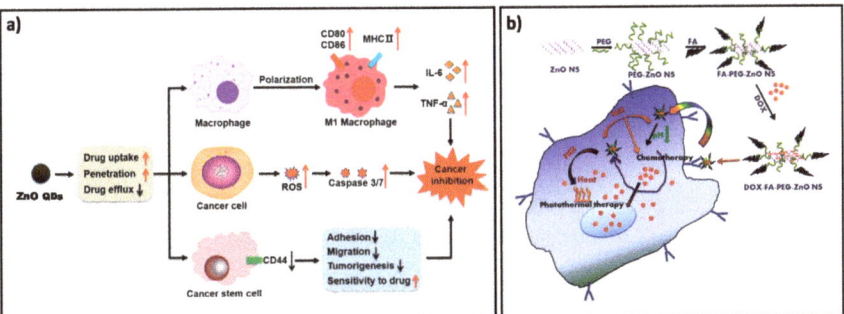

Figure 2. (a) Scheme of the multiple proposed effects of ZnO QDs as a multi-functional antitumour treatment. Reproduced with permission from [44]. American Chemical Society, 2017; (b) Scheme of the combined mechanism of action of DOX-FA-ZnO NS for breast carcinoma therapy. Reproduced with permission from [48]. Elsevier, 2017.

Bahadur et al. demonstrated that the application of ultrasound irradiation in ZD QDs can be used for on-demand pulsatile release of DOX molecules [45]. To increase the selectivity and luminescence of the nanocarrier, Zhu et al. designed Mg ZnD QDs functionalized with Folic Acid (FA) and studied their toxicity in HeLa cells [46]. Pathak et al. also used FA as targeting agent to synthesize a new ZnO hollow-nanocarrier containing paclitaxel as model drug. Initially, they suspended carbon spheres in zinc acetate solution and added ammonia to form a zinc hydroxide layer onto the surface of the carbon spheres. After that, the carbon was removed by pyrolysis, giving rise to the hollow ZnO spheres. Then, the NPs were loaded with paclitaxel and functionalized with FA. The effectiveness of the nanosystem was successful both in vitro by producing cytotoxicity with breast cancer cells and in vivo by reducing MDA-MB-231 xenograft tumours in nude mice [47]. FA modified zinc oxide nanosheets (Ns) were also proposed by Kannan et al. as a chemo-photothermal device for breast cancer therapy. The experiments showed that the combination of both therapies (chemotherapy and photothermal therapy) resulted in higher percentages of cell death than either of them separately. In addition, in vitro and in vivo experiments showed no adverse effect or toxicity on blood stream. In Figure 2b, a scheme of the combined mechanism of action of these nanosheets is presented [48]. As explained in the Introduction, the irradiation of ZnO QDs with UV light increases the production of ROS, enhancing the antitumour capacity of the QDs. Several groups have studied this effect in tumour cells with pure particles [12], particles doped with other ions [33], in combination with different antitumour agents [49,50] or using an aptamer as targeting agent [51].

2.2. ZnO Core Nanocomposites

Several nanocomposites based on ZnO QDs in combination with other materials to obtain new nanocomposites with synergist theranostic effects have also been reported. For example, H. Möhwald et al. synthesized ZnO QDs with a polymeric shell, coordinated with Gd^{3+} ions and adsorbed DOX to create a versatile ZnO-Gd-DOX nanodevice. It was a bifunctional probe for both in vitro fluorescent and in vivo animal imaging, due to the strong red emission of ZnO-Gd-DOX in the range of 600–800 nm and magnetic resonance imaging (MRI) contrast due to Gd^{3+} ions, which were immobilized onto the ZnO surface through proper coordination with carboxyl groups of the polymer. This rendered an outstanding relaxivity for MRI. Most importantly, these nanomaterials also demonstrated a very promising antitumour activity. BxPC-3 tumour-bearing nude mice were injected with different agents to study chemotherapy efficacy. As shown in Figure 3a, the tumour in the control continued growing, while those treated with DOX, Doxil or ZnO-Gd-DOX QDs remained more or less the same. H&E (hematoxylin and eosin) staining of tumour slices (Figure 3b) showed that the cells in the control retained their normal membrane and nuclear structures, and the cells treated with DOX or Doxil were damaged partly, while almost all cells were severely destroyed after ZnO-Gd-DOX treatment. This result confirmed that ZnO-Gd-DOX QDs performed better than the other agents. Finally, they also observed that ZnO-Gd-DOX QDs had no detectable toxic side effects to mice, and the whole QDs could be biodegraded and excreted from the mice body [34].

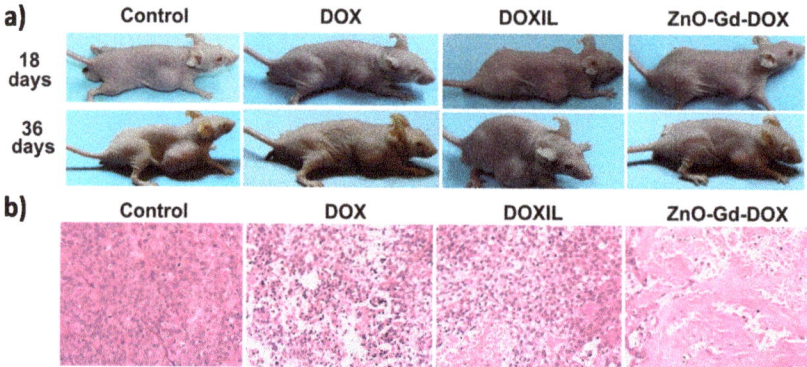

Figure 3. (a) Images of BxPC-3 tumour-bearing nude mice after 18 and 36 days under different treatments; (b) H&E staining of tumour slices after 36 days of treatments by different agents. Reproduced with permission from [34]. American Chemical Society, 2016.

Qu et al. combined the advantages of mesoporous silica nanoreactors, DOX, FA, and ZnO QDs to develop a drug carrier effectively protected from non-specific degradation (ZnO-DOX@F-mSiO$_2$-FA). They demonstrated that the mesoporous silica shell protected the ZnO-DOX device from non-specific protein degradation, while retained its sensitivity to pH-responsiveness. To perform the experiments, HeLa and HEK 293T cells, which are, respectively, positive and negative for folate receptor, were selected as model cells. After the treatment with ZnO-DOX@mSiO$_2$-FA, a clear increase in positive annexin V-FITC HeLa cells was observed when compared with control ones. The results also showed an effective targeting due to the presence of the folic acid as ZnO-DOX@mSiO$_2$-FA presented selective toxic capacity toward HeLa cells [52]. Cauda et al. designed new lipid-coated ZnO nanocrystals (NCs) to achieve a better stability in biological samples. Their results showed that lipid-coated ZnO NCs presented stable colloidal dispersions in cell culture medium and simulated human plasma for 25 days. However, after being suspended, the pristine and amine-functionalized NCs quickly aggregated, remaining stable for less than an hour. Even though internalization of lipid-shielded ZnO NCs in HeLa cells was higher compared to the other samples, its toxicity was lower, showing a lower toxicity/particle ratio [53].

2.3. ZnO Coated Nanodevices

Due to its high biocompatibility, in recent years ZnO has been used as a coating for other types of QDs that present more toxicity. In comparison with the pegylation that also provides stability and biocompatibility, the use of ZnO as a coating provides added values such as luminescent properties or ROS production, among others, that cannot be achieved by PEG functionalization. For instance, H. Danafar et al. recently reported a new system, TiO$_2$@ZnO NPs, where ZnO is used to coat mesoporous TiO$_2$ QDs. They loaded the mesopores with curcumin (Cur) and studied their pH-dependent in vitro anticancer effect against human epithelial colorectal adenocarcinoma cells. The cytotoxic capacity was also compared with a similar system that contained graphene oxide

(GO), (which has probe to be quite efficient in the treatment of cancer of colon) [62,63] as final layer, TiO$_2$@ZnO–GO. Opposite to what was expected, the presence of GO did not increase toxicity. Proof of this is that TiO$_2$@ZnO showed higher killing capability against Caco-2 cancer cell than the ones that contained GO. Therefore, there was reduced toxicity of ZnO nanoparticles [26]. Authors suggested that effect could be because of the presence of ZnO nanoprecipitation on the tumour cells [64,65]. In 2015, Wang et al. reported two different systems based on iron oxide QDs coated with ZnO and sensitive to magnetic and microwave radiations. In both cases, the Fe$_3$O$_4$ core functioned for magnetic targeting, allowing them, with the help of an external magnet, to concentrate the NPs in the desire tissue, while the ZnO shell acted as a microwave absorber that facilitated the release of the drug due to an increase in temperature. The first system consisted of Fe$_3$O$_4$@ZnO@mGd$_2$O$_3$:Eu@P(NIPAm-co-MAA) NPs used as drug carrier of the cytotoxic etoposide (VP-16). The mesoporous Gd$_2$O$_3$:Eu shells acted as drug nanocarrier, being the poly[(N-isopropylacrylamide)-co-(methacrylic acid)] polymer (P(NIPAm-co-MAA)) the temperature sensitive caps that responded to microwave application. The experiments demonstrated that the ZnO shells effectively absorbed and converted microwave irradiation to heat; as a result, P(NIPAm-co-MAA) contracts, unblocking the mesopores and triggering the release of about 81.7% of the entrapped VP16 drug within 10 h [54]. The second nanoprobes were core–shell structured β-CD-Fe$_3$O$_4$@ZnO:Er^{3+},Yb^{3+} nanoparticles and their scheme of synthesis is shown in Figure 4a. In this system, the drug was stored in the inert cavity of the β-cyclodextrins (β-CD) due to hydrophobic interactions. The ZnO shell doped with Er^{3+} and Yb^{3+} not only acted as microwave absorber that produd a thermal response (similar to the previous one) but also provided fluorescence imaging for in vitro detection. The data showed that β-CD-Fe$_3$O$_4$@ZnO:Er^{3+},Yb^{3+} NPs were able to transform the microwaves into localized internal heating, allowing the release of the drug in a control manner by selecting the microwave exposure time and the number of cycles applied (Figure 4b). The MTT assay showed that the NPs had a strong targeting effect producing high rates of tumour cell death almost without affecting healthy ones [55].

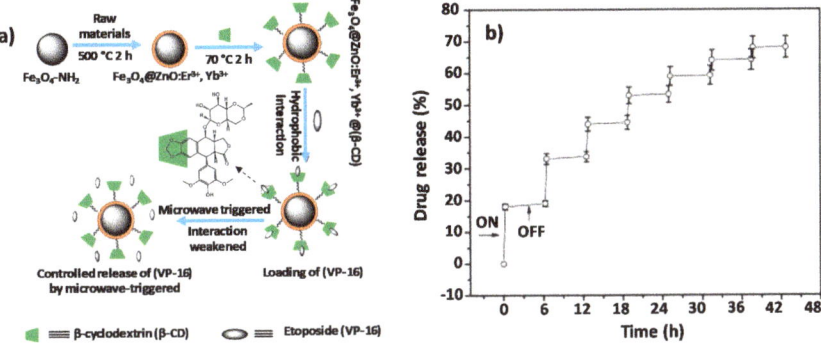

Figure 4. (a) Synthesis scheme and mechanism of action of Fe$_3$O$_4$@ZnO:Er^{3+},Yb^{3+}@β-CD nano-composites; (b) Graph of VP-16 release from the Fe$_3$O$_4$@ZnO:Er^{3+},Yb^{3+}@(β-CD)–(VP-16) depending on the number of microwave cycles applied. Reproduced with permission from [55]. Elsevier, 2015.

2.4. ZnO QDs as Pore Caps

ZnO QDs are sensitive to pH and can be synthesized at different sizes. These two features make them great candidates to act as "gatekeeper" of the pores of bigger systems, allowing to enclose in its interior different drugs that will only be released under acidic tumour conditions after dissolution of the QDs. Based on this idea, the mesoporous silica nanoparticles with a large load capacity and a pore diameter around 2.5 nm seem to be the perfect combination [66–68]. In fact, several groups have proposed different systems based on the combination of these two components, mesoporous silica nanoparticles as storage material and ZnO QDs as cap of the pores. It was in 2011 when G. Zhu et al. reported this combination for the first time. They synthesized MSN with amino groups inside the pores and carboxylic acid groups outside and loaded them with DOX. Finally, they used amino functionalized ZnO nanolids (Nls) to close the pores through amide coupling with 1-ethyl-3-(3-dimethylaminopropyl) carbodiimide (EDC). The stimulus-responsive capacity of the DOX-loaded ZnO MSNs were studied at different pH values. At physiological pH (7.4), negligible DOX release was observed; however, at pH 5.0m a fast DOX release was observed due to the dissolution of the ZnO Nls (Figure 5a). As can be seen in Figure 5b, the viability studies demonstrated that ZnO MSNs@DOX led to high death rates in HeLa cells at very low concentrations (6.25 µg/mL) [56].

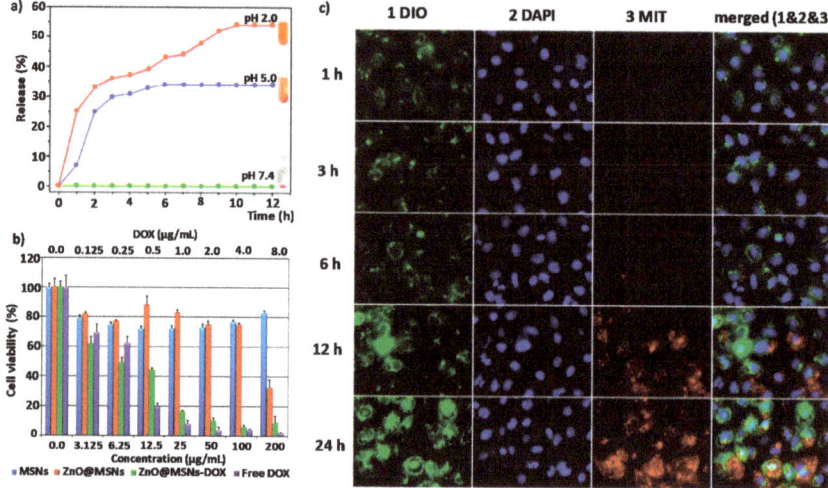

Figure 5. (a) DOX release profiles from ZnO@MSNsDOX at 3 different pH values (7.4, 5.0, and 2.0); (b) In vitro viability of HeLa cells in the presence of COOHMSNs, ZnO@MSNs, ZnO@MSNsDOX, and free DOX. Reproduced with permission from [56]. American Chemical Society, 2011; (c) Confocal microscopy images taken at different times of A549 cells incubated with the MIT-loaded, ZnO-gated MCNs. Cell membranes were stained in green, cell nuclei were stained in blue and released drugs wre presented red. Reproduced with permission from [60]. Elsevier, 2016.

Two years later, the same group presented a new system whose novelty lay in the fact that ZnO QDs were also used as chelate forming agents of a second hydrophobic drug (curcumin) that was significantly loaded onto the surface of the ZnO Nls. Cell viability experiments confirmed that the combination of both drugs presented a high cytotoxic effect even for small device concentrations such as ~3 µg/mL [57]. In 2015, H. Zhang et al. came out with a multifunctional nanotheranostic agent with lanthanide-doped upconverting nanoparticles (UCNPs) (NaYF$_4$: 20%Yb^{3+}, 2%Er^{3+}/NaGdF$_4$: 2%Yb^{3+}) as the core for the UCL/CT/MRI trimodality imaging. In this system, called UCNPs@mSiO$_2$-ZnO, the core was cover with a mesoporous silica layer, loaded with a drug and finally pH-sensitive ZnO QDs were used to cap the pores. UCNPs@mSiO$_2$-ZnO demonstrated great abilities to be employed as contrast agents for tri-modal imaging, both in vitro and in vivo. The cytotoxic effect of the loaded NPs was studied against HeLa cell and the results showed higher therapeutic effectiveness than that obtained with the free DOX [58]. Lei Sun et al. prepared a new protease/redox/pH triple-sensitive delivery nanodevice through the combination of fluorescent ZnO QDs, oxidized glutathione (GSSG) and amino-functionalized pSiO$_2$ NPs. In that case, GSSG was linked to the surface of NH$_2$-functionalized pSiO$_2$ NPs through amido bonds. Then, ZnO QDs were also covalently attached to GSSG acting as fluorescence probes and caps of the pores. In that system, the release of the amoxicillin cargo was susceptible to being triggered by three different stimuli: (i) the degradation of ZnO QDs in acid media characteristic of tumour environments; (ii) the break of the GSSG linkers through amido bonds by proteases; or (iii) the break of the GSSG linkers through disulphide bonds by the presence of reduced glutathione (GSH). Three different release experiments were performed to study the effect of each of the stimuli separately. It was observed that the amoxicillin release was slower because of GSH than that in the acidic pH conditions, which authors attribute to the presence ZnO QDs hindering GSH from entering the mesoporous structure to break the disulphide bonds. However, this dependency showed that the addition of protease K was higher, achieving around 66% of cargo release in only 6 h. The authors expected that this triple-stimuli responsive nanodevice had great applications for antitumour therapy [59]. One year after, in 2017 the same research group enhanced the GSH response of the nanodevice by anchoring the ZnO QDs through cystine (Cys) molecules and studied the in vitro cytotoxic effect of the system against HepG$_2$ cells. After incubation for 48 h without drug cargo, the cell viability was about 96.6%, even at a high concentration of 500 µg/mL, which demonstrates that both L-pSiO$_2$/Cys and L-pSiO$_2$/Cys/ZnO presented good biocompatibility. However, when the same experiment was carried out, the DOX-loaded L-pSiO2/Cys nanoparticles displayed a concentration-dependent cytotoxicity significantly higher than that observed for free DOX [25]. Mesoporous carbon nanoparticles (MCNs) present similar characteristics to those described for MSNs in terms of mechanical and structural properties. That is why X. Du et al. reported a new carboxylated MCNs were ZnO QDs were covalently linked via dual amide bonds to minimize premature release of mitoxantrone (MIT) drug. The toxicity of the system itself was studied in vitro in A549 cells showing low toxicity after 48 h of incubation. However, when ZnO-gated MCNs were loaded with MIT the tumour killing capability was clearly increased to values of about 65% even at small drug concentrations such us 0.066 ng/cell. As can be seen in Figure 5c, the release of the drug was monitored by confocal microscopy, showing that MIT was exclusively released inside the cells without premature release in the extracellular media. This proved that the release is due to the dissolution of the ZnO QDs by a decrease in the pH inside the cell [60].

2.5. ZnO QDs That Provides an Added Value to Other Systems

Microgels are three-dimensional networks which have various applications due to their facile fabrication, their drug loading capacity and the possibility to functionalize them to obtain stimuli responsive effect. There are two strategies to achieve these stimuli responsive microgels: utilizing sensitive polymers and designing microgels with degradable crosslinkage. The latter strategy was used by J. Feng et al. to design ZnO@Dextran microgels loaded with DOX, were amino-modified ZnO QDs acted as pH-sensitive crosslinkers to carboxymethyl dextran (CMD)(Figure 6a–c). Since ZnO QDs are dissolvable at pH lower than 5.5, degradation of the ZnO@Dextran microgels at different pH values (7.4, 5.0, and 3.0) was investigated. As shown in Figure 6d, the digital photos were taken after vortexing of the microgels in different buffers for several minutes. It was observed that the hybrid microgels dissolved completely at pH 5.0 and 3.0, yielding a clear solution but in the case of pH 7.4 a non-transparent solution could be seen, indicating that most of the microgel particles remained intact. To evaluate tumour therapeutic effect of the DOX-loaded ZnO@Dextran microgels, the cytotoxicities of ZnO@Dextran/DOX and the two control materials (ZnO@Dextran microgels and free DOX) were investigated by the authors in Hela cells by MTT assays. By an unusual procedure, the cells were incubated with one of those materials under pH condition of 5.0 and 7.4 for 24 h, respectively. As shown in Figure 6e, ZnO@Dextran/DOX showed more pronounced cytotoxic effects on Hela cells at pH 5.0 than 7.4 [61].

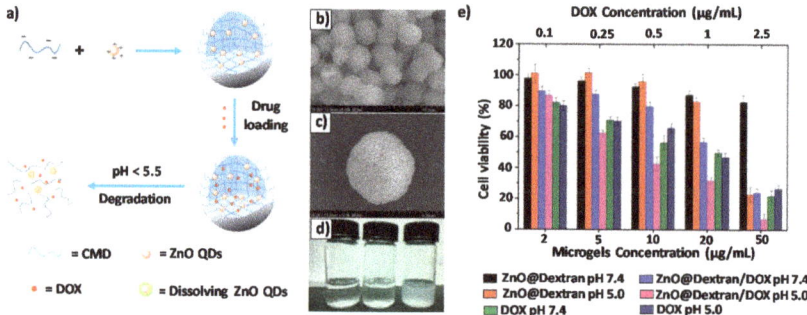

Figure 6. (a) Schematic representation of fabrication and degradation process of the ZnO@Dextran microgels; (**b,c**) Two different magnification SEM photographs of ZnO@Dextran microgels; (**d**) Digital photos of ZnO@Dextran microgels after incubation at pH 3.0 (left), 5.0 (middle) and 7.4 (right); (**e**) Cell viabilities of HeLa cells after being incubated with different samples under different conditions. Reproduced with permission from [61]. John Wiley and Sons, 2018.

3. ZnO Nanoplatforms for Bacterial Infection

The increase in the quality of life has been associated with an increase in the longevity of the population and, therefore, to the number of fragile and immunocompromised people, who are candidates to suffer a bacterial infection. In addition, despite aseptic surgical techniques, the number of surgical interventions that are practiced each year is increasing, making it the appearance of a bacterial infection associated with implants even more possible. Thus, bacterial infections have attracted great attention from both the medical community and public worldwide [69]. There are several obstacles that hinder the effectiveness of bacterial infection treatments: (i) the lack of early stage detection methods [70], which in clinic are currently limited to indirect imaging modalities; (ii) the overuse of antibiotics favouring the appearance of bacteria resistant to multiple drugs, leading to antibiotic ineffectiveness [71]; (iii) biofilm formation, which has an extracellular matrix that makes it resistant to antibiotics [72]; amd (iv) intracellular infections, whose difficulty lies in the wide variety of mechanisms used by bacteria to ensure their survival such as inhibition of the phago/lysosome fusion, resistance to lysosomal enzymes attack, etc. [73].

Thus, it would be necessary to develop clinical imaging tools able to selectively attack invasive bacteria and/or biofilm, allowing early detection [74] and that at the same time can release some antibiotics or produce some therapeutic effect in the infected area [73]. Luminescent ZnO QDs, with high surface-to-volume ratio, capacity of ROS production and a strong surface chemistry, meet all the necessary requirements to adequately treat bacterial infection [75]. Besides, recent studies evidence that NPs can present antibacterial capacity without having any side effect on human cells, that is, showing selective toxicity [76]. This selectivity might be because of differences in the action mechanisms in bacteria and mammal cells. For bacteria, the most recognized NPs action mechanisms so far are ROS and free radical generation, bacterial cells adhesion and penetration, biofilm penetration and changes on bacteria metabolic activity and gene expression [77].

It has become clear that biofilm-grown cells exhibit different properties than planktonic bacteria, therefore nanodevices with different properties need to be use for the treatment of each one [78].

3.1. ZnO Nanoplatforms for Planktonic Bacteria Treatment

ZnO QDs have proven their effectiveness against a broad spectrum of pathogenic microorganisms, even though the response they produce in Gram-positive/-negative studies are different, probably due to differences in the membrane composition and in the intracellular antioxidant content of both types of bacteria [7]. It has also been reported that the antimicrobial capacity of the ZnO clearly depends on the size of the QDs, i.e., with smaller sizes being more effective [79].

The antibacterial effect of ZnO QDs on *Campylobacter jejuni* was investigated by X. Shi et al. for inhibition and inactivation of cell growth. In this study, the effect of ZnO QDs on the integrity of the bacterial membrane and on the expression of different bacteria genes was studied. Ethidium monoazide (EMA) is a fluorescent label able to bind DNA and inhibit PCR amplification. However, the bacterial membrane is impermeable to EMA so the union only take place if the membrane has been damaged. EMA-PCR experiments were carried out and a clear reduction on DNA amplification was observed for the samples treated with ZnO QDs. Besides an increase expression of ahpC and katA, two common genes that express oxidative stress, was also observed. These PCR results suggested that the antibacterial mechanism of ZnO nanoparticles in Campylobacter was most likely due to disruption of the cell membrane and oxidative stress [80]. I. Ahmad et al. performed a comparative experiment to study the antibacterial activity of ZnO QDs against different types of bacteria, including some Gram-negative (*Vibrio cholera*, *Campylobacter jejuni* and *Escherichia*) and Gram-positive (methicillin resistant *Staphylococcus aureus*) studies. After 4 h of ZnO QDs exposure, a clear decrease in bacterial viability was observed in all cases, being especially effective against *C. jejuni* where bacterial death amounted to 65%. On the other hand, *V. cholerae* were the least affected bacteria seeing its population reduced by only 27% [75]. S. Arastoo et al. synthesized Ag and ZnO QDs and mixed them to study their antibacterial effect against *Mycobacterium Tuberculosis* (Mtb) growth inside macrophages. The toxicity of the QDs in THP-1 leukemic cells was also studied. Results showed that the 8ZnO/2Ag ratio was the most effective since it had high antibacterial capacity against Mtb both in vivo and in vitro but without affecting the cellular viability of THP-1 cells [81]. In 2016, R. Jalal et al. used different techniques (fluorescence and scanning electron microscopy, flow cytometry and DNA extraction) to elucidate the mechanism of action of ZnO QDs administered alone and in combination with two antibiotics ciprofloxacin and ceftazidime. The authors demonstrated that sub-inhibitory concentrations of ZnO QDs were enough to significantly increase the uptake and antibacterial capacity of both antibiotics.

The effect of light on increasing ROS production for the treatment of bacteria had also been studied. J. Gupta and D. Bahadur reported the antibacterial and anticancer activity effect of visible light irradiation on Cu substituted ZnO nanoassemblies (Cu-ZnO NAs). First, different concentrations of Cu were studied to optimize the ROS production with Cu_5-ZnO being the most efficient material. The data showed that the NAs themselves had antibacterial activity. Although irradiation with visible light alone had a negligible effect on cell viability, the combination of light with the Cu5-ZnO NAs produced 100% of bacterial death after 30 min of light application [82].

Opposite to what has been said so far, VanEpps et al. have recently published an article that casts serious doubts on ROS production being responsible for the antibacterial efficacy of ZnO QDs. They based their conclusions on experiments made with two different types of ZnO nanoparticles against *S. aureus* bacteria. Bacteria were exposed to both types of NPs and concentrations of H_2O_2 that presented an equivalent amount of bacterial killing capacity. Additionality, the same experiment was carried out but in presence of the antioxidant N-acetylcysteine (NAC). Results showed a clear recovery of those bacterial colonies treated with H_2O_2 but no changes in the ones treated with ZnO laying bare that the decrease in bacteria population was not due to ROS production (Figure 7).

Studies on the effect of ZnO NPs on gene expression were consistent with that as oxidative stress genes were down regulated. In addition, the effect on anaerobic carbohydrate metabolism and energetics with upregulation of the UMP biosynthesis pathway was also proposed as the reason for ZnO NPs antibacterial activity [83].

As in the case of cancer, the use of ZnO nanosystems for the treatment of bacterial infections is not only limited to QDs themselves, but ZnO QDs can also be used as a component of more complex systems. In 2015, Parkin et al. incorporated crystal violet and zinc oxide nanoparticles (CVZnO) into medical grade polyurethane polymers to synthesize new surfaces with antibacterial activity. The antibacterial capacity of these surfaces after light irradiation proved to be lethal for the two types of bacteria studied: E. coli and S. aureus, for Gram-negative and Gram-positive, respectively [84]. One year later, Gu et al. developed a fluorescent nano-probe MPA@ZnO-PEP by the combination of silica stabilized ZnO QDs (ZnO@SiO$_2$) with PEP, a peptide fragment for bacteria targeting, and MPA, a near infrared dye. This fluorescent nanodevice proved to be a potent tool to bacteria detection both in vitro and in vivo since it allowed to differentiate between sterile inflammation and bacterial infection. To increase the antibacterial capacity, vancomycin (Van) was also incorporated to the nanoplatform to form the so-called MPA/Van@ZnO-PEP. Antibacterial (S. aureus and B. subtilis) test were performed, showing that for B. subtilis a concentration of 1mg was enough to produce bacterial inhibition, needing 2 mg to achieve the same effect in S. aureus. In both cases, it was a much lower concentration than that required for free Van. Furthermore, authors wanted to test the efficacy of the nanosystem against antibiotic resistant bacteria. Therefore, Van was substituted by methicillin (Met) and the antibacterial capability of MPA/Met@ZnO-PEP was studied against to a S. aureus strain which was resistant to this antibiotic (MRSA). While bacteria incubated with free Met or the unloaded nanoplatform exhibited a viability comparable to the control, those incubated with MPA/Met@ZnO-PEP reduced their viability to a 60% for a concentration of 64 µg/mL of NPs [85].

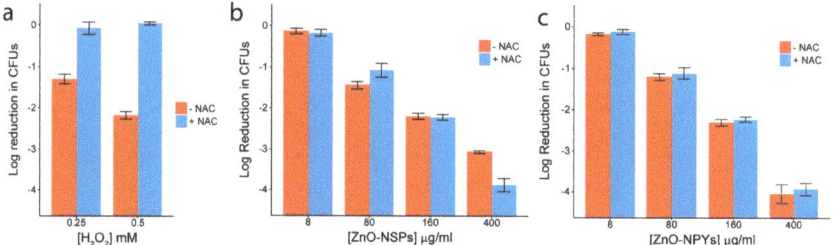

Figure 7. Reduction in colony counts (from 5 × 10^7) observed after exposure to increasing concentrations of: (a) H$_2$O$_2$; (b) ZnO-NSPs; or (c) ZnO-NPYs, with and without 50 mM NAC. Reproduced with permission from [83]. Royal Society of Chemistry, 2018.

3.2. ZnO Nanoplatforms for Biofilm Treatment

Biofilms are defined as complex communities of microorganisms that grow embedded in a self-produced protective extracellular matrix made of polysaccharides, DNA and proteins. The main problem related with biofilm formation is that at this point bacteria becomes antibiotic resistant and more tolerant to disinfectant chemicals or to the response of the immune system, increasing the probability to degenerate into chronic infections [72]. Clinical observations and experimental studies clearly indicate that a treatment exclusively based in the administration of antibiotics is usually insufficient to eliminate the biofilm. Hence, to effectively eradicate biofilm infections, new, more complex strategies such as the use of ZnO nanodevices are needed.

Juršėnas et al. focused their research on the effect of unmodified ZnO nanorods (NRs) against strongly resistant bio-films after 405 nm light excitation. Therefore, different bacteria biofilms were grown onto surfaces covered by ZnO NRs. Then some of them were exposed to light while the others remained in darkness. The results showed that neither the NRs nor the irradiation with light presented by themselves any type of toxicity against planktonic bacteria or biofilms (see Figure 8).

However, for those samples coated with ZnO NRs and subject to light, the biofilm was substantially reduced. Besides, it was observed that inactivation of biofilm growth was strongly dependent on light dose. Moreover, different degrees of effectiveness were observed for the same treatment depending on the type of bacteria used, with L. monocytogenes biofilms being the most susceptible population and E. faecalis the most resistance one [86].

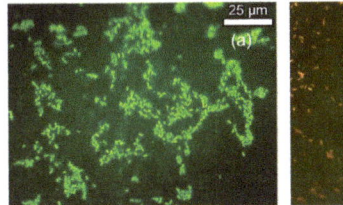 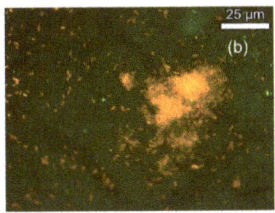 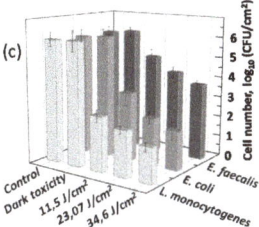

Figure 8. Fluorescence micrographs of *E. coli* biofilms grown up onto ZnO NRs surfaces: (**a**) without; or (**b**) with light treatment. Alive cells were stained in green and dead cells in red; (**c**) Light-dose dependence cytotoxic effect on *E. faecalis* MSCL 302, *L. monocytogenes* ATCL3C 7644 and *E. coli* O157:H7 biofilms grown up onto ZnO NRs coated surfaces. Non-illuminated biofilms grown on plastic surfaces were used as control and effect the NRs in absence of light was expressed as dark toxicity. Reproduced with permission from [86].

3.3. ZnO Nanoplatforms for Planktonic and Biofilm Treatment

The ultimate treatment would be one that involve a simultaneous attack of the planktonic bacteria and the undesired bacterial biofilm. Recently, several systems based on ZnO QDs have been shown to fulfil this double functionality. An example of this was the research carried out by A.A. Al-Khedhairy et al. about the effect of ZnO QDs against bacterial strain *Pseudomonas*. For the study of the effect on bacteria, several plates were cultured with increasing QD concentrations, observing a maximum inhibition at 100 mg/mL of ZnO QDs. To follow the inhibition of biofilm formation, crystal violet dye was employed as stain. The results showed a significant inhibition of the biofilm formation at 50 and 100 mg/mL of ZnO QDs [87]. J. Lee et al. also used *Pseudomonas aeruginosa* as bacteria but in this case the effect of thirty-six metal ions was studied. Among all of them, zinc ions and ZnO QDs were able to efficiently reduce *P. aeruginosa* biofilm as well as the production of different bacterial growth signals [88]. Similarly, A. Jamalli et al. studied the effect of ZnO QDs in the biofilm formation and the antigen 43 expression (an important surface protein in *E. coli* which is encoded by flu gene). The authors observed that ZnO QDs produced inhibitory effects on biofilm formation in uropathogenic *Escherichia coli* (UPEC) isolates. Results also showed that concentrations of QDs lower than the amount needed to inhibit biofilm formation were nevertheless enough to significantly decrease the expression of the flu gene in UPEC [89].

Chakrabarti et al. studied the effect of ZnO QDs on two biotypes of cholera bacteria (classical and El Tor) observing greater efficiency for El Tor strain both in biofilm and in planktonic forms. The authors were not completely sure about the reason for the differences in susceptibility. However, according to their suggestion, it might be due to some differences in membrane structure or gene expression between the two biotypes. Results showed that ZnO QDs produced ROS that damaged bacterial membrane and substantially modified their morphology. Authors also tested the antibacterial capacity of the QDs in cholera toxin (CT) mouse models. As can be seen in Figure 9a, controls presented strong fluid accumulation in the loops and therefore a big organ distention and diarrheal symptoms. Fluid accumulation values equal to 0.2 or higher are considered a positive result in diarrhoea. Figure 9b shows that a single administration of ZnO QDs with CT was enough to considerably reduce intestinal fluid accumulation. The synergist effect of the combine administration of ZnO QDs and kanamycin was also studied achieving a considerably increased killing capability of 85–87% killing compared with 50–70% of the QDs alone [90].

Another important contribution of ZnO against infection is its use as coating on implant materials. In 2016, Rose et al. reported the effect of three different ZnO QDs structures (spheres, plates and pyramids) in planktonic growth and biofilm formation after surface coating. Several bacterial types were used, and planktonic growth experiments revealed that *S. epidermidis* and *S. aureus* (Gram-positive bacteria) presented dose dependent reduction for all ZnO structures. However, *K. pneumonia* and *E. coli* (Gram-negative bacteria) were not affected by the presence of ZnO QDs up to 667 µg/mL, probably because of differences in bacterial surface hydrophobicity.

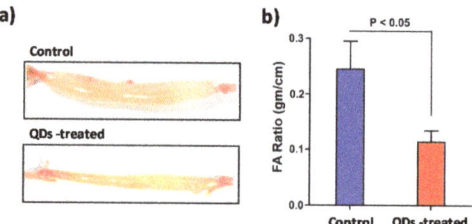

Figure 9. (**a**) Image of isolated mouse intestinal loop 6 h after the injection of 1 µg of CT and 1.25 µg µL^{-1} of ZnO QDs or only CT in the control one; (**b**) Fluid accumulation (FA) ratio after 6 h of injection; (n = 10 mice with 15–20 loops studied per group). Reproduced with permission from [90]. Elsevier, 2016.

Biofilm formation experiments on polystyrene ZnO QDs coated surfaces demonstrated no significant differences related with the particle morphology. Nevertheless, notable differences were observed in terms of the type of bacteria studied allowing a reduction in biofilm formation only for Gram-positive bacteria [91]. De Fátima Montemor et al. studied the antibacterial and antibiofilm effect of nano-and micro-sized ZnO coatings against MRSA. Which found that only the nano-sized coating was able to reduce biofilm formation. Additionally, the combination effect of ZnO coating followed by the addition of different antibiotics in sub-inhibitory concentrations was also analysed. The results showed a greater antibacterial efficacy due to the combine treatment with gentamicin, no differences when adding trimethoprim and even an infection worsening in the presence of rifampicin, ciprofloxacin, and vancomycin [92]. Recently, Wang et al. reported a new method to coat implant surfaces based on the combination of two different ZnO structures named as ZnO nanorods—nanoslices hierarchical structure (NHS). Authors modified the surface of two commonly used implant materials (titanium and tantalum) with NHS and each of the structures separately and studied their antibacterial capacity against *E. coli* and *S. aureus* under different times. Results proved that ZnO nanoslices rapidly released (48 h) allowing to kill bacteria in early stages. On the other hand, ZnO nanorods that presented higher stability needed around two weeks to present a bacteria killing capability. ZnO NHS, which combined both elements presented a two-stage antibacterial effect. The antibacterial efficacy of ZnO NHS was also studied by mice test in vivo, showing not only good results onto the implant surfaces but in the surrounding areas where an effective sterilization was also observed [93].

However, despite the numerous advantages of the antibacterial character of ZnO QDs, a poor control of their residues can have negative effects on soil-dwelling microorganisms responsible for numerous activities of great impact on the soil quality such as plant protection, biodegradation, ecological balance and nutrient recycling [94].

All recent advances made for the treatment of bacterial infection with ZnO nanoplatforms are summarized in Table 2.

Table 2. ZnO nanoplatforms for theranostic bacterial infection.

Type of Bacteria Used [a]	Type of Device [b]	Responsive Phenomena [c]	Drug/Antibiotic [d]	Reference
C. jejuni	ZnO QDs	-	-	[80]
EPEC, C. jejuni, V. Cholerae, MRSA	ZnO QDs	-	-	[75]
THP-1, M. tuberculosis	ZnO QDs + Ag QDs	-	-	[81]
A. baumannii	ZnO QDs	-	Coadministered Cip, Cef	[95]
E. coli	Cu-ZnO NAs	Visible light	-	[82]
MRSA	ZnO-NPYs, ZnO QDs	-	-	[83]
S. aureus, E. coli	CVZnO polyurethane surface	White light	Loaded CV	[84]
S. aureus, B. subtilis, MRSA, S. aureus or B. subtilis infected mouse	MPA@ZnO-PEP	-	Loaded Met, Van	[85]
MSCL 302, ATC$_{L3}$C 7644, O157:H7	ZnO NRs	UV light	-	[86]
P. aeruginosa	ZnO QDs	-	-	[87]
P. aeruginosa	ZnO QDs	-	-	[88]
E. coli	ZnO QDs	-	-	[89]
V. cholerae, Mouse intestinal loop	ZnO QDs	-	Coadministered kanamycin	[90]
S. epidermidis, S. aureus, K. pneumonia, E. coli	ZnO (spheres, plates, pyramids)	-	-	[91]
MRSA	Nano- and micro-sized ZnO coatings	-	Coadministered, Gent, Trim, Rif, Cip, Van	[92]
S. aureus, E. coli, mice	NHS, ZnO NRs, ZnO NSs	-	-	[93]

[a] EPEC: Enteropathogenic *Escherichia coli*; *C. jejuni*: *Campylobacter jejuni*; *V. cholerae*: *Vibrio cholerae*; MRSA: Methicillin resistant *Staphylococcus aureus*; THP-1: Human monocytic cell; *M. tuberculosis*: *Mycobacterium tuberculosis*; *A. baumannii*: *Acinetobacter baumannii*; *S. aureus*: *Staphylococcus aureus*; *S. epidermidis*: *Staphylococcus epidermidis*; *K. pneumonia*: *Klebsiella pneumonia*; MSCL 302: *Enterococcus faecalis*; ATC$_{L3}$C 7644: *Listeria monocytogenes*; O157:H7: *E. coli*; *P. aeruginosa*: *Pseudomonas aeruginosa*. [b] ZnO QDs: Zinc oxide quantum dots; Ag QDs: Silver quantum dots; Cu-ZnO NAs: Cu substituted ZnO nanoassemblies; ZnO-NPYs: ZnO nanopyramids with hexagonal base; ZnO QDs CVZnO: Crystal violet and zinc oxide nanoparticles; MPA@ZnO-PEP: Silica stabilized ZnO quantum dots with an antibacterial peptide fragment (UBI$_{29-41}$) and a near infrared dye MPA (derived from hydrophilic indocyanine green); NHS: ZnO nanorods–nanoslices hierarchical structure; NRs: Nanorods; NSs: Nanoslices. [c] UV: Ultra violet; [d] Cip: Ciprofloxacin; Cef: Ceftazidime; Met: Methicillin; Van: Vancomycin; Gent: gentamicin; Trim: trimethoprim; and Rif: rifampicin.

4. Antifungal Capacity of ZnO Nanoplatforms

The antimicrobial power of the ZnO QDs is not limited to bacteria but also to other types of microorganisms such as fungi. The advantage of these materials is that their effectiveness can be applied to treat different problems caused by microorganisms such as, infections, diseases, biocontamination and corrosion.

Growth of fungal pathogens is one of the main problems in agriculture and usually causes high economic losses to farmers [96]. Different groups have reported the antifungal efficacy of ZnO QDs against plant pathogens. M. Lin et al. observed that ZnO QDs at concentration higher than 3 mmol/L could markedly reduce the growth of *B. cinerea* getting even better results against *P. expansum* (Figure 10) [97].

The antifungal effect of ZnO QDs has also been proven against other plant pathogens [98–100].

Another important problem related with microorganism and especially with fungi is the biological colonization of stone heritage what gives rise to its biodegradation. Recently, P. Quintana et al. studied the antifungal and photocatalytic properties of ZnO, MgO, and Zn/Mg Oxide QDs for the protection of calcareous stone heritage. Results showed that the growth inhibition of *Penicillium oxalicum*, *Pestalotiopsis maculans*, *Paraconiothyrium* sp., and *Aspergillus niger* fungi achieved by Mg$_{1-x}$Zn$_x$O QDs was higher than that obtained by either of the two pure oxide QDs. The photocatalytic activity observed was also higher in the case of the mixed oxides compare with the pure ones [101].

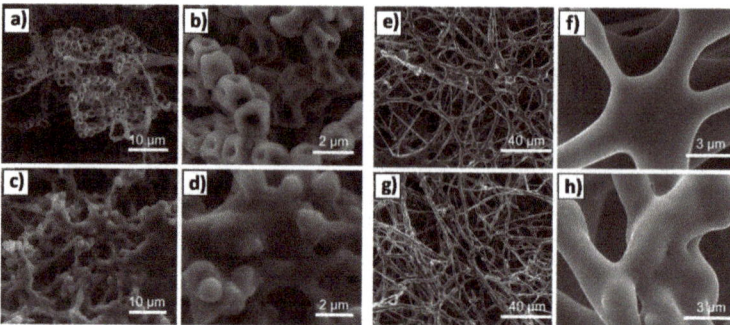

Figure 10. SEM images of (**a–d**) *Penicillium expansum* and (**e–h**) *Botrytis cinerea* without (a,b,e,f) or with (c,d,g,h) the treatment of ZnO QDs suspension. Reproduced with permission from [97]. Elsevier, 2011.

The infection in humans due to the action of opportunistic fungi is also a relevant health problem nowadays. N.O. Jasim reported the effect of ZnO QDs against two different opportunistic fungi (*A. fumigatus*, *C. albicans*) and observed a significant decreased ($p \geq 0.05$) in the radial growth of the fungus at different concentrations and an increase in the inhibitory effect when increasing the period of incubation [102]. To increase the antifungal effectiveness against *Candida albicans*, Ding et al. designed ZnO QDs coated by chitosan (CS) and functionalized with linoleic acid (LiA). Chitosan has been reported to present antimicrobial activity because of the presence of positive charges that react with cellular DNA of some bacteria [103]. Additionally, LiA with Trans-11 and Cis-9 arrangement is one of the effective acids on fungal growth [104]. The authors studied the effect of CS-LiA QDs compared with that produced by fluconazole (a potent antifungal drug). Results showed that CS-LiA QDs could inhibit the growth of fluconazole-resistant clinical strains at similar concentrations to fluconazole, but the inhibition percent of biofilm formation, for nanoparticles was greater than that of fluconazole [105]. The use of ZnO QDs to synthesize antifungal filters [106] or as antifungal bioglasses [107] for medical implants or surgical equipment has also been reported as a probe into the versatility of these QDs in the fight against microorganisms. All ZnO based nanoplatforms for antifungal treatment cited in this manuscript are summarized in the following Table 3.

Table 3. ZnO nanoplatforms for antifungal treatment.

Type of Fungi Used [a]	Type of Device [b]	Responsive Phenomena [c]	Reference
B. cinerea, P. expansum	ZnO QDs	-	[97]
A. saloni, S. rolfii	ZnO QDs	-	[98]
R. stolonifera, A. flavus, A. nidulans, T. harzianum	ZnO QDs	-	[99]
E. salmonicolor	ZnO QDs	-	[100]
A. niger, P. oxalicum, Paraconiothyrium sp., P. maculans	Zn/Mg Oxide QDs	UV light	[101]
A. fumigatus, C. albicans	ZnO QDs	-	[102]
C. albicans	CS-LiA ZnO QDs	-	[105]
R. stolonifera, P. expansum	ZnO QDs	-	[106]
C. krusei	ZnO QDs	-	[107]

[a] B. cinerea: Botrytis cinerea; P. expansum: Penicillium expansum; A. saloni: Alternaria saloni; S. rolfii: Sclerotium rolfsii; A. flavus: Aspergillus flavus; A. nidulans: Aspergillus nidulans; T. harzianum: Trichoderma harzianum; E. salmonicolor: Erythricium salmonicolor; A. niger: Aspergillus niger; P. oxalicum: Penicillium oxalicum; P. maculans: Pestalotiopsis maculans; A. fumigatus: Aspergillus fumigatus; C. albicans: Candida albicans; C. krusei: Candida krusei. [b] ZnO QDs: Zinc oxide quantum dots; CS-LiA ZnO QDs: ZnO QDs coated by chitosan (CS) and functionalized with linoleic acid (LiA). [c] UV: Ultra violet.

5. ZnO Nanoplatforms for Diabetes Treatment

According with the National Diabetes Statistics Report of 2017 [108], 30.3 million people of all ages, or 9.4% of the U.S. population, had diabetes in 2015 and, unfortunately, the number of people suffering from this disease continues to grow over time. These figures make diabetes one of the greatest diseases facing our society today. In 1934, when it was observed that insulin crystal contained zinc [109], researchers began to believe that Zn, insulin and diabetes could be intimately related. Nowadays, it is known that Zn interacts with some elements of insulin signalling pathway and therefore affects the glucose metabolism. Besides, other aspects of diabetes disease such as β-cell function, glucose homeostasis, insulin action, and diabetes pathogenesis are also influenced by the presence of this ion [110]. As the degradation of ZnO results in the release of Zn^{2+} ions in the media, the number of studies about the incidence of ZnO QDs in the treatment of diabetes has increased in recent years. Umrani and Paknikar tested the antidiabetic effect of ZnO QDs in rats and observed that after oral administration the glucose tolerance was improved. Other factors were also enhanced such us a reduction in non-esterified fatty acids, triglycerides and glucose in blood and an increase in serum insulin [111]. In 2015 Asri-Rezaie et al. performed a comparative studied of the antidiabetic activity and toxic effects of ZnO QDs and zinc sulphate ($ZnSO_4$) in diabetic rats. It was observed that those treated with ZnO showed greater antidiabetic activity compared to $ZnSO_4$. However, severely elicited oxidative stress particularly at higher doses was also observed [112]. The efficacy of ZnO QDs in attenuating pancreatic damage in a rat model of treptozotocin-induced diabetes was also settled one year later by Kandeel et al. [113]. Other comparative studies including ZnO QDs for antidiabetic studies have been performed [114]. In 2016, Abu-Risha et al. went a step further and studied the antidiabetic effect of a co-administered treatment based on ZnO QDs and Vildagliptin, a standard antidiabetic drug. As a result, the recovery of the function and structure of β cells and a synergistic effect on the therapy of Type-2 diabetes between this two components was observed [115]. S.N. Kale et al. also explored the combination effect between ZnO and drugs but in this case the antidiabetic agent was conjugated to the surface of the QDs. Red Sandalwood (RSW), a potent natural extract anti-diabetic agent was chosen as model drug. Results showed that ZnO-RSW QDs was more effective against the crude murine pancreatic glucosidase than any of the two elements (RSW and ZnO QDs) that composed it administered separately [116]. All ZnO based nanoplatforms for diabetes treatment cited in this are summarized in Table 4.

Table 4. ZnO nanoplatforms for diabetes treatment.

Model Used	Type of Device [a]	Drug/Antibiotic [b]	Reference
Rats	ZnO QDs	-	[111]
Rats	ZnO QDs, $ZnSO_4$	-	[112]
Rats	ZnO QDs	-	[113]
Rats	ZnO, CeO_2, Ag QDs, MC	-	[114]
Rats	ZnO QDs	Coadministered Vildagliptin	[115]
Murine Pancreatic and Small Intestinal Extracts	ZnO QDs	Conjugated RSW	[116]

[a] ZnO QDs: Zinc oxide quantum dots; CeO_2 QDs: cerium oxide quantum dots; Ag QDs: silver quantum dots; MC: *Momordica charantia*. [b] RSW: Red Sandalwood.

6. ZnO Nanoplatforms with Anti-Inflammatory Properties

The beneficial health effects of both elemental zinc and its salts have been known for a long time, which is why for several hundred years they have been widely used for therapeutic purposes, including the anti-inflammatory effectiveness in several common inflammatory dermatoses and for wounds or ulcers [117]. Since the appearance of nanoparticles, and considering these beneficial properties of zinc ions, the anti-inflammatory capacity of ZnO QDs have also been studied. H. Alenius et al. investigated the response produced by topically administration of nano-sized ZnO (nZnO) in the mouse model of atopic dermatitis (AD) and compared these outcomes to those induced by bulk-sized ZnO (bZnO). Their experiments clearly demonstrated that nZnO efficiently reduced the thickness of the skin in the allergic environment compared with the effect produced by bZnO. While topical application of bZnO favoured macrophage infiltration, the recruiting of CD8+ and CD4+ T cells in allergic skin was strongly inhibited by nZnO treatment. These results demonstrated that topical nZnO treatment had great effects in reducing skin inflammation as consequence of allergen exposition [118].

The anti-inflammatory capacity of ZnO QDs is not limited to skin problems but has also proven to be very effective for other inflammatory diseases. Shin et al. studied the anti-inflammatory properties of ZnO in RAW 264.7 murine macrophage cells by measuring their effect in pro-inflammatory mediators. As can be seen in Figure 11, ZnO QDs clearly reduced inflammation and showed a dose-dependent effect in the suppression of different protein expressions (COX-2, iNOS, TNF-α and interleukins-1β and -6) and mRNA [119].

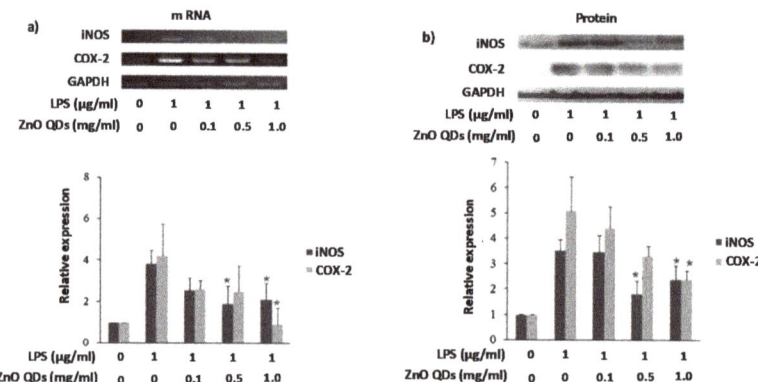

Figure 11. (a) Determination of iNOS and COX-2 mRNA levels by RT-PCR; (b) Determination of iNOS and COX-2 protein levels by Western blot (mean ± SD of n = 3 and * $p < 0.05$ versus LPS alone). Reproduced with permission from [119]. Elsevier, 2015.

Inflammatory bowel diseases are widespread inflammatory diseases that cause debilitating health problems including cancer. In their research, Feng et al. showed that ZnO QDs treatment had markedly dose-dependent effects on the remission of dextran sulphate sodium influenced ulcerative colitis in mice. They also demonstrated that the antioxidant and anti-inflammatory abilities of ZnO QDs were related to their capacity to suppress ROS and malondialdehyde production; increase GSH level; suppress pro-inflammatory cytokines IL-1β and TNF-α and myeloperoxidase (MPO) [120]. Throughout this manuscript the ability of ZnO QDs to produce ROS has been highlighted. However, the anti-inflammatory properties of these particles are now related to their ability to eliminate them. Although this may seem contradictory, the truth is that the same properties (large surface area and high catalytic capacity) that favour the generation of ROS in environments where they are not present is probably responsible for suppressing them in environments where they are very abundant. Ulcerative colitis tissues had 10- to 100-fold increased ROS, and in this case, the antioxidant activity of ZnO QDs might be the result of electron density transfer from the oxygen to the odd electron placed on outer orbits of oxygen in $O_2^{\bullet-}$ and $OH^\bullet$ radicals [121]. Selected ZnO based nanoplatforms with anti-inflammatory properties are presented in Table 5.

Table 5. ZnO nanoplatforms with anti-inflammatory properties.

Model Used [a]	Type of Device [b]	Reference
AD model mouse	nZnO, bZnO	[118]
RAW 264.7	ZnO QDs	[119]
mice	ZnO QDs	[120]

[a] AD: Atopic dermatitis; RAW 264.7: Murine macrophage cells. [b] nZnO: nano-sized ZnO; bZnO: bulk-sized ZnO; ZnO QDs: Zinc oxide quantum dots.

7. ZnO Nanoplatforms for Wound Healing

The lack of zinc production delaying wound healing is a fact well known in clinic. Already in 1990, the effect of topically applied zinc on leg ulcer healing and its effect on some mechanisms in wound healing was studied using standardized animal models [122]. After that, several clinical and experimental studies were performed with elemental Zn and zinc oxide [123]. Results showed that treatments based in topical ZnO application produced several benefits in re-epithelialization, wound healing, infections and ulceration among others [124]. Currently, researchers are mostly focusing their efforts on the design of new materials for wound healing that incorporate the antibacterial and healing effects of ZnO QDs. R. Jayakumar et al. synthesized a microporous chitosan hydrogel based on ZnO composite bandages (CZBs) that presented great flexibility. In addition, ZnO QDs were incorporated into a hydrogel made of chitosan. Results showed that these nanocomposite bandages presented antibacterial activity and great effects on enhancing blood clotting and swelling capacity. Infiltration and cell attachment studies were performed showing a clear penetration and nanocomposite attachment of the cells. Furthermore, as can be seen in Figure 12, the in vivo test in Sprague-Dawley rats proved the efficacy of these bandages in wound healing as well as their beneficial effects on collagen deposition and better re-epithelialization [125].

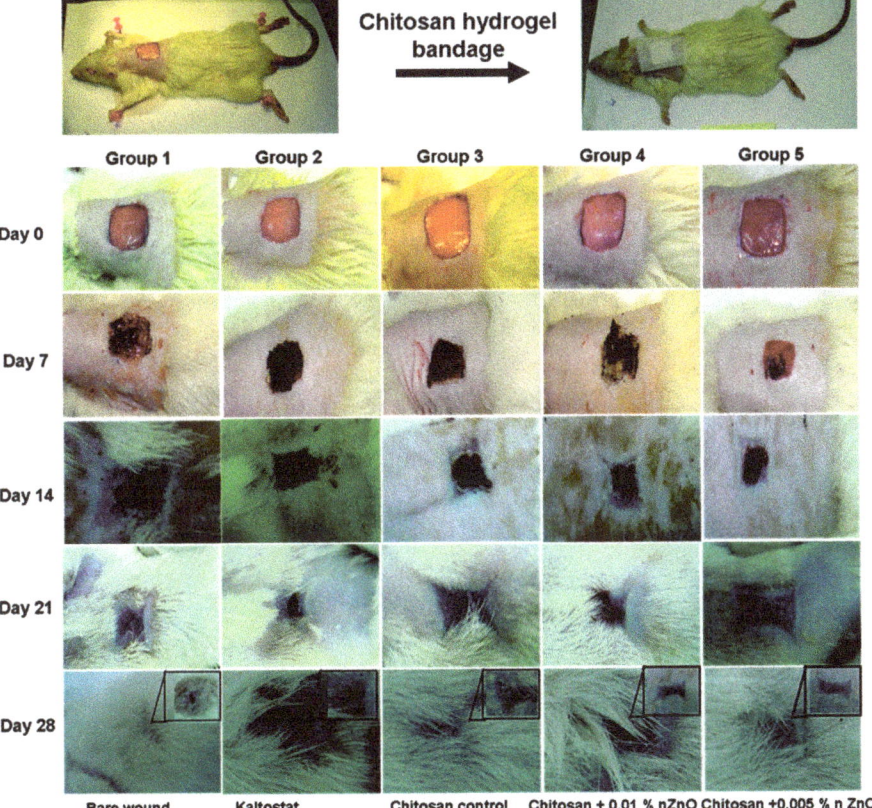

Figure 12. In vivo images of the wound healing process in Sprague−Dawley rats. Reproduced with permission from [125]. American Chemical Society, 2012.

Manuja et al. synthesized ZnO QDs-loaded-sodium alginate-gum acacia hydrogels (SAGA-ZnO QDs) by bounding the aldehyde and hydroxyl groups of gluteradehyde and sodium alginate polymer, respectively. Results showed that hydrogels were biocompatible in peripheral blood fibroblast/mononuclear cells and could produce inhibition of *Pseudomonas aerigunosa* and *Bacillus cereus*. SAGA-ZnO QDs hydrogels also presented a great healing capacity in sheep fibroblast cells even at low concentrations [126].

Due to its excellent properties including biocompatibility, haemostatic, bacteriostatic and healing capacity, chitosan has been extensively used in biomedicine [127]. In 2017, two research groups reported two different composites based on chitosan and containing ZnO QDs to combine antibacterial and wound healing properties in the same material. Denkbas et al. proposed the use of scaffolds containing chitosan/silk sericin (CHT/SS) in combination with lauric acid (LA) or ZnO QDs for wound dressing applications. Although both combinations CHT/SS/ZnO QDs and CHT/SS/LA had antimicrobial effect against *E. coli* and *S. aureus*, the presence of LA proved to be clearly more efficient for both bacteria. Besides, these nanocomposites were able to improve the attachment, proliferation and growth of HaCaT cells, presenting no secondary effects [128]. The second composite was proposed by Bezerra et al. and consisted in chitosan-based films that contained chondroitin 4-Sulphate, gelatin and zinc oxide nanoparticles. The wound healing capacity of the system was evaluated in rat models with full-thickness excision. After six days, results showed a significant decrease (14–35% more) of the wound compared with that obtained in control ones [129]. Selected ZnO based nanoplatforms for wound healing are summarized in Table 6.

Table 6. ZnO nanoplatforms for wound healing.

Model Used [a]	Type of Device [b]	Reference
nHDF cells, SD rats	CZBs	[125]
PBMC, sheep fibroblast cells	SAGA-ZnO QDs hydrogels	[126]
HaCaT	CHT/SS/ZnO QDs, CHT/SS/LA	[128]
rats	CHT/gel/C4S/ZnO films	[129]

[a] nHDF: Normal human dermal fibroblasts; SD: Sprague-Dawley; PBMC: Peripheral blood mononuclear cells; HaCaT: Aneuploid immortal keratinocyte cell. [b] CZBs: Microporous chitosan hydrogel/nano zinc oxide composite bandages; SAGA-ZnO QDs hydrogels; CHT/SS/ZnO QDs: Chitosan/silk sericin scaffolds combined with ZnO QDs; CHT/SS/LA: Chitosan/silk sericin scaffolds combined with lauric acid; CHT/gel/C4S/ZnO films: chitosan-based films containing gelatin, chondroitin 4-Sulphate and ZnO, respectively.

8. Conclusions and Future Outlook

We demonstrate in this article of bibliographic review that zinc oxide nanoparticles present unique properties: (i) luminescence; (ii) Zn^{2+} ions release in aqueous media, especially in acidic conditions; (iii) ROS generation, mainly after being irradiated with UV light; (iv) versatile surface chemistry; and (v) an easy and economical synthetic process that allows an accurate control of the size and shape of the particles. All these factors make this nanomaterial a very versatile system, by itself or combined with other elements. This proves that ZnO nanodevices have applications in virtually all fields of science, especially in biomedicine. In this review, we report the latest advances derived from their used in the treatment of cancer and diabetes; their antibacterial, antifungal and anti-inflammatory efficacy; and their wound healing capacity. Despite all the advances in the understanding of ZnO nanostructures mechanism of action and biological effects, there is still a lack of knowledge on this subject, especially on what the long-term effects are. Therefore, a greater knowledge is necessary to assess whether the already multiple known advantages of the ZnO outweigh the potential risks. In addition, nowadays, reproducibility in the NPs synthesis process is still a challenge to be achieved. This, together with the lack of a standardized protocol (concentrations used, times of action, etc.) that allows comparing the different studies, makes it even more complex to obtain a complete knowledge of the NPs effects and therefore their clinical translation. However, in view of the facts, the opinion of the authors is that this is only the beginning and that, in the near future, ZnO based nanodevices will not only will achieve clinical states but other novel theranostic applications will also be proposed for their use.

Acknowledgments: M. Vallet-Regí acknowledges the European Research Council, ERC-2015-AdG (VERDI), Proposal No. 694160 and Ministerio de Economía y Competitividad (MINECO) (MAT2015-64831-R grant). M. Martínez-Carmona also thanks Irish Research Council for the Postdoctoral fellowship.

Conflicts of Interest: The authors declare no conflict of interest.

References

1. Jia, Z.; Misra, R.D.K. Tunable ZnO quantum dots for bioimaging: synthesis and photoluminescence. *Mater. Technol.* **2013**, *28*, 221–227. [CrossRef]
2. Resch-Genger, U.; Grabolle, M.; Cavaliere-Jaricot, S.; Nitschke, R.; Nann, T. Quantum dots versus organic dyes as fluorescent labels. *Nat. Methods* **2008**, *5*, 763–775. [CrossRef] [PubMed]
3. Volokitin, Y.; Sinzig, J.; de Jongh, L.J.; Schmid, G.; Vargaftik, M.N.; Moiseevi, I.I. Quantum-size effects in the thermodynamic properties of metallic nanoparticles. *Nature* **1996**, *384*, 621–623. [CrossRef]
4. Asok, A.; Gandhi, M.N.; Kulkarni, A.R. Enhanced visible photoluminescence in ZnO quantum dots by promotion of oxygen vacancy formation. *Nanoscale* **2012**, *4*, 4943. [CrossRef] [PubMed]
5. Zhang, Z.-Y.; Xiong, H.-M. Photoluminescent ZnO Nanoparticles and Their Biological Applications. *Materials* **2015**, *8*, 3101–3127. [CrossRef]
6. Vandebriel, R.J.; De Jong, W.H. A review of mammalian toxicity of ZnO nanoparticles. *Nanotechnol. Sci. Appl.* **2012**, *5*, 61–71. [CrossRef] [PubMed]
7. Espitia, P.J.P.; Soares, N.F.F.; Coimbra, J.S.; dos de Andrade, R.; Cruz, R.S.; Medeiros, E.A.A. Zinc Oxide Nanoparticles: Synthesis, Antimicrobial Activity and Food Packaging Applications. *Food Bioprocess Technol.* **2012**, *5*, 1447–1464. [CrossRef]
8. Zhang, L.; Yin, L.; Wang, C.; Qi, Y.; Xiang, D. Origin of Visible Photoluminescence of ZnO Quantum Dots: Defect-Dependent and Size-Dependent. *J. Phys. Chem. C* **2010**, *114*, 9651–9658. [CrossRef]
9. Xiong, H.-M.; Shchukin, D.G.; Möhwald, H.; Xu, Y.; Xia, Y.-Y. Sonochemical Synthesis of Highly Luminescent Zinc Oxide Nanoparticles Doped with Magnesium(II). *Angew. Chem. Int. Ed.* **2009**, *48*, 2727–2731. [CrossRef] [PubMed]
10. Manaia, E.B.; Kaminski, R.C.K.; Caetano, B.L.; Briois, V.; Chiavacci, L.A.; Bourgaux, C. Surface modified Mg-doped ZnO QDs for biological imaging. *Eur. J. Nanomed.* **2015**, *7*, 109–120. [CrossRef]
11. Fan, Z.; Lu, J.G. Zinc Oxide Nanostructures: Synthesis and Properties. *J. Nanosci. Nanotechnol.* **2005**, *5*, 1561–1573. [CrossRef] [PubMed]
12. Zhang, H.; Chen, B.; Jiang, H.; Wang, C.; Wang, H.; Wang, X. A strategy for ZnO nanorod mediated multi-mode cancer treatment. *Biomaterials* **2011**, *32*, 1906–1914. [CrossRef] [PubMed]
13. Driscoll, K.E.; Howard, B.W.; Carter, J.M.; Janssen, Y.M.W.; Mossman, B.T.; Isfort, R.J. *Mitochondrial-Derived Oxidants and Quartz Activation of Chemokine Gene Expression BT—Biological Reactive Intermediates VI: Chemical and Biological Mechanisms in Susceptibility to and Prevention of Environmental Diseases*; Dansette, P.M., Snyder, R., Delaforge, M., Gibson, G.G., Greim, H., Jollow, D.J., Monks, T.J., Sipes, I.G., Eds.; Springer US: Boston, MA, USA, 2001; pp. 489–496, ISBN 978-1-4615-0667-6.
14. Wilson, M.R.; Lightbody, J.H.; Donaldson, K.; Sales, J.; Stone, V. Interactions between ultrafine particles and transition metals in vivo and in vitro. *Toxicol. Appl. Pharmacol.* **2002**, *184*, 172–179. [CrossRef] [PubMed]
15. Rasmussen, J.W.; Martinez, E.; Louka, P.; Wingett, D.G. Zinc Oxide Nanoparticles for Selective Destruction of Tumor Cells and Potential for Drug Delivery Applications. *NIH Public Accsess* **2011**, *7*, 1063–1077. [CrossRef] [PubMed]
16. Kim, Y.J.; Yu, M.; Park, H.O.; Yang, S.I. Comparative study of cytotoxicity, oxidative stress and genotoxicity induced by silica nanomaterials in human neuronal cell line. *Mol. Cell. Toxicol.* **2010**, *6*, 337–344. [CrossRef]
17. Ostrovsky, S.; Kazimirsky, G.; Gedanken, A.; Brodie, C. Selective cytotoxic effect of ZnO nanoparticles on glioma cells. *Nano Res.* **2009**, *2*, 882–890. [CrossRef]
18. Liou, M.-Y.; Storz, P. *Reactive Oxygen Species in Cancer*; Taylor & Francis: Oxfordshire, UK, 2010; Volume 44, ISBN 1071576100.

19. Lepot, N.; Van Bael, M.K.; Van den Rul, H.; D'Haen, J.; Peeters, R.; Franco, D.; Mullens, J. Synthesis of ZnO nanorods from aqueous solution. *Mater. Lett.* **2007**, *61*, 2624–2627. [CrossRef]
20. Xi, Y.; Hu, C.G.; Han, X.Y.; Xiong, Y.F.; Gao, P.X.; Liu, G.B. Hydrothermal synthesis of ZnO nanobelts and gas sensitivity property. *Solid State Commun.* **2007**, *141*, 506–509. [CrossRef]
21. Tripathi, R.M.; Bhadwal, A.S.; Gupta, R.K.; Singh, P.; Shrivastav, A.; Shrivastav, B.R. ZnO nanoflowers: Novel biogenic synthesis and enhanced photocatalytic activity. *J. Photochem. Photobiol. B Biol.* **2014**, *141*, 288–295. [CrossRef] [PubMed]
22. Samadipakchin, P.; Mortaheb, H.R.; Zolfaghari, A. ZnO nanotubes: Preparation and photocatalytic performance evaluation. *J. Photochem. Photobiol. A Chem.* **2017**, *337*, 91–99. [CrossRef]
23. Liu, W.C.; Cai, W. Synthesis and characterization of ZnO nanorings with ZnO nanowires array aligned at the inner surface without catalyst. *J. Cryst. Growth* **2008**, *310*, 843–846. [CrossRef]
24. Cho, E.B.; Volkov, D.O.; Sokolov, I. Ultrabright fluorescent mesoporous silica nanoparticles. *Small* **2010**, *6*, 2314–2319. [CrossRef] [PubMed]
25. Zhang, X.; Zhao, Y.; Cao, L.; Sun, L. Fabrication of degradable lemon-like porous silica nanospheres for pH/redox-responsive drug release. *Sens. Actuators B Chem.* **2018**, *257*, 105–115. [CrossRef]
26. Zamani, M.; Rostami, M.; Aghajanzadeh, M.; Kheiri Manjili, H.; Rostamizadeh, K.; Danafar, H. Mesoporous titanium dioxide@ zinc oxide–graphene oxide nanocarriers for colon-specific drug delivery. *J. Mater. Sci.* **2018**, *53*, 1634–1645. [CrossRef]
27. Kawabata, Y.; Wada, K.; Nakatani, M.; Yamada, S.; Onoue, S. Formulation design for poorly water-soluble drugs based on biopharmaceutics classification system: Basic approaches and practical applications. *Int. J. Pharm.* **2011**, *420*, 1–10. [CrossRef] [PubMed]
28. Hu, J.; Johnston, K.P.; Williams, R.O. Nanoparticle Engineering Processes for Enhancing the Dissolution Rates of Poorly Water Soluble Drugs. *Drug Dev. Ind. Pharm.* **2004**, *30*, 233–245. [CrossRef] [PubMed]
29. Martínez-Carmona, M.; Lozano, D.; Colilla, M.; Vallet-Regí, M. Selective topotecan delivery to cancer cells by targeted pH-sensitive mesoporous silica nanoparticles. *RSC Adv.* **2016**, *6*, 50923–50932. [CrossRef]
30. Manuscript, A.; Barriers, G.M. Oral Drug Delivery with Polymeric Nanoparticles: The Gastrointestinal Mucus Barriers. *NIH Public Access* **2013**, *64*, 557–570. [CrossRef]
31. Bahrami, B.; Hojjat-Farsangi, M.; Mohammadi, H.; Anvari, E.; Ghalamfarsa, G.; Yousefi, M.; Jadidi-Niaragh, F. Nanoparticles and targeted drug delivery in cancer therapy. *Immunol. Lett.* **2017**, *190*, 64–83. [CrossRef] [PubMed]
32. Kou, L.; Bhutia, Y.D.; Yao, Q.; He, Z.; Sun, J.; Ganapathy, V. Transporter-Guided Delivery of Nanoparticles to Improve Drug Permeation across Cellular Barriers and Drug Exposure to Selective Cell Types. *Front. Pharmacol.* **2018**, *9*, 27. [CrossRef] [PubMed]
33. Ghaemi, B.; Mashinchian, O.; Mousavi, T.; Karimi, R.; Kharrazi, S.; Amani, A. Harnessing the Cancer Radiation Therapy by Lanthanide-Doped Zinc Oxide Based Theranostic Nanoparticles. *ACS Appl. Mater. Interfaces* **2016**, *8*, 3123–3134. [CrossRef] [PubMed]
34. Ye, D.X.; Ma, Y.Y.; Zhao, W.; Cao, H.M.; Kong, J.L.; Xiong, H.M.; Möhwald, H. ZnO-Based Nanoplatforms for Labeling and Treatment of Mouse Tumors without Detectable Toxic Side Effects. *ACS Nano* **2016**, *10*, 4294–4300. [CrossRef] [PubMed]
35. Król, A.; Pomastowski, P.; Rafińska, K.; Railean-Plugaru, V.; Buszewski, B. Zinc oxide nanoparticles: Synthesis, antiseptic activity and toxicity mechanism. *Adv. Colloid Interface Sci.* **2017**, *249*, 37–52. [CrossRef] [PubMed]
36. Mirzaei, H.; Darroudi, M. Zinc oxide nanoparticles: Biological synthesis and biomedical applications. *Ceram. Int.* **2017**, *43*, 907–914. [CrossRef]
37. Singh, A.; Singh, N.B.; Afzal, S.; Singh, T.; Hussain, I. Zinc oxide nanoparticles: a review of their biological synthesis, antimicrobial activity, uptake, translocation and biotransformation in plants. *J. Mater. Sci.* **2018**, *53*, 185–201. [CrossRef]
38. Naveed Ul Haq, A.; Nadhman, A.; Ullah, I.; Mustafa, G.; Yasinzai, M.; Khan, I. Synthesis Approaches of Zinc Oxide Nanoparticles: The Dilemma of Ecotoxicity. *J. Nanomater.* **2017**, *2017*, 1–14. [CrossRef]
39. Ludi, B.; Niederberger, M. Zinc oxide nanoparticles: chemical mechanisms and classical and non-classical crystallization. *Dalt. Trans.* **2013**, *42*, 12554–12568. [CrossRef] [PubMed]

40. David, C.A.; Galceran, J.; Rey-Castro, C.; Puy, J.; Companys, E.; Salvador, J.; Monné, J.; Wallace, R.; Vakourov, A. Dissolution kinetics and solubility of ZnO nanoparticles followed by AGNES. *J. Phys. Chem. C* **2012**, *116*, 11758–11767. [CrossRef]
41. Bisht, G.; Rayamajhi, S. ZnO Nanoparticles: A Promising Anticancer Agent. *Nanobiomedicine* **2016**, *3*, 9. [CrossRef]
42. Sharma, H.; Kumar, K.; Choudhary, C.; Mishra, P.K.; Vaidya, B. Development and characterization of metal oxide nanoparticles for the delivery of anticancer drug. *Artif. Cells Nanomed. Biotechnol.* **2016**, *44*, 672–679. [CrossRef] [PubMed]
43. Liu, J.; Ma, X.; Jin, S.; Xue, X.; Zhang, C.; Wei, T.; Guo, W.; Liang, X.J. Zinc Oxide Nanoparticles as Adjuvant to Facilitate Doxorubicin Intracellular Accumulation and Visualize pH-Responsive Release for Overcoming Drug Resistance. *Mol. Pharm.* **2016**, *13*, 1723–1730. [CrossRef] [PubMed]
44. Wang, J.; Lee, J.S.; Kim, D.; Zhu, L. Exploration of Zinc Oxide Nanoparticles as a Multitarget and Multifunctional Anticancer Nanomedicine. *ACS Appl. Mater. Interfaces* **2017**, *9*, 39971–39984. [CrossRef] [PubMed]
45. Barick, K.C.; Nigam, S.; Bahadur, D. Nanoscale assembly of mesoporous ZnO: A potential drug carrier. *J. Mater. Chem.* **2010**, *20*, 6446. [CrossRef]
46. Muhammad, F.; Guo, M.; Guo, Y.; Qi, W.; Qu, F.; Sun, F.; Zhao, H.; Zhu, G. Acid degradable ZnO quantum dots as a platform for targeted delivery of an anticancer drug. *J. Mater. Chem.* **2011**, *21*, 13406. [CrossRef]
47. Puvvada, N.; Rajput, S.; Prashanth Kumar, B.; Sarkar, S.; Konar, S.; Brunt, K.R.; Rao, R.R.; Mazumdar, A.; Das, S.K.; Basu, R.; Fisher, P.B.; Mandal, M.; Pathak, A. Novel ZnO hollow-nanocarriers containing paclitaxel targeting folate-receptors in a malignant pH-microenvironment for effective monitoring and promoting breast tumor regression. *Sci. Rep.* **2015**, *5*, 1–15. [CrossRef] [PubMed]
48. Vimala, K.; Shanthi, K.; Sundarraj, S.; Kannan, S. Synergistic effect of chemo-photothermal for breast cancer therapy using folic acid (FA) modified zinc oxide nanosheet. *J. Colloid Interface Sci.* **2017**, *488*, 92–108. [CrossRef] [PubMed]
49. Deng, Y.; Zhang, H. The synergistic effect and mechanism of doxorubicin-ZnO nanocomplexes as a multimodal agent integrating diverse anticancer therapeutics. *Int. J. Nanomed.* **2013**, *8*, 1835–1841. [CrossRef] [PubMed]
50. Hackenberg, S.; Scherzed, A.; Harnisch, W.; Froelich, K.; Ginzkey, C.; Koehler, C.; Hagen, R.; Kleinsasser, N. Antitumor activity of photo-stimulated zinc oxide nanoparticles combined with paclitaxel or cisplatin in HNSCC cell lines. *J. Photochem. Photobiol. B Biol.* **2012**, *114*, 87–93. [CrossRef] [PubMed]
51. Han, Z.; Wang, X.; Heng, C.; Han, Q.; Cai, S.; Li, J.; Qi, C.; Liang, W.; Yang, R.; Wang, C. Synergistically enhanced photocatalytic and chemotherapeutic effects of aptamer-functionalized ZnO nanoparticles towards cancer cells. *Phys. Chem. Chem. Phys.* **2015**, *17*, 21576–21582. [CrossRef] [PubMed]
52. Yan, Z.; Zhao, A.; Liu, X.; Ren, J.; Qu, X. A pH-switched mesoporous nanoreactor for synergetic therapy. *Nano Res.* **2017**, *10*, 1651–1661. [CrossRef]
53. Dumontel, B.; Canta, M.; Engelke, H.; Chiodoni, A.; Racca, L.; Ancona, A.; Limongi, T.; Canavese, G.; Cauda, V. Enhanced biostability and cellular uptake of zinc oxide nanocrystals shielded with a phospholipid bilayer. *J. Mater. Chem. B* **2017**, *5*, 8799–8813. [CrossRef] [PubMed]
54. Qiu, H.; Cui, B.; Zhao, W.; Chen, P.; Peng, H.; Wang, Y. A novel microwave stimulus remote controlled anticancer drug release system based on Fe_3O_4@ZnO@mGd_2O_3:Eu@P(NIPAm-co-MAA) multifunctional nanocarriers. *J. Mater. Chem. B* **2015**, *3*, 6919–6927. [CrossRef]
55. Peng, H.; Cui, B.; Li, G.; Wang, Y.; Li, N.; Chang, Z.; Wang, Y. A multifunctional β-CD-modified Fe3O4ZnO:Er3 +,Yb3 +nanocarrier for antitumor drug delivery and microwave-triggered drug release. *Mater. Sci. Eng. C* **2015**, *46*, 253–263. [CrossRef] [PubMed]
56. Muhammad, F.; Guo, M.; Qi, W.; Sun, F.; Wang, A.; Guo, Y.; Zhu, G. PH-triggered controlled drug release from mesoporous silica nanoparticles via intracelluar dissolution of ZnO nanolids. *J. Am. Chem. Soc.* **2011**, *133*, 8778–8781. [CrossRef] [PubMed]
57. Muhammad, F.; Wang, A.; Guo, M.; Zhao, J.; Qi, W.; Yingjie, G.; Gu, J.; Zhu, G. PH dictates the release of hydrophobic drug cocktail from mesoporous nanoarchitecture. *ACS Appl. Mater. Interfaces* **2013**, *5*, 11828–11835. [CrossRef] [PubMed]

58. Wang, Y.; Song, S.; Liu, J.; Liu, D.; Zhang, H. ZnO-functionalized upconverting nanotheranostic agent: Multi-modality imaging-guided chemotherapy with on-demand drug release triggered by pH. *Angew. Chem. Int. Ed.* **2015**, *54*, 536–540. [CrossRef]
59. Qiu, L.; Zhao, Y.; Li, B.; Wang, Z.; Cao, L.; Sun, L. Triple-stimuli (protease/redox/pH) sensitive porous silica nanocarriers for drug delivery. *Sens. Actuators B Chem.* **2017**, *240*, 1066–1074. [CrossRef]
60. Huang, X.; Wu, S.; Du, X. Gated mesoporous carbon nanoparticles as drug delivery system for stimuli-responsive controlled release. *Carbon N. Y.* **2016**, *101*, 135–142. [CrossRef]
61. Zhang, J.; Chen, L.; Chen, J.; Wu, D.; Feng, J. Dextran microgels loaded with ZnO QDs: pH-triggered degradation under acidic conditions. *J. Appl. Polym. Sci.* **2018**, *135*, 1–6. [CrossRef]
62. Chen, G.-Y.; Pang, D.W.-P.; Hwang, S.-M.; Tuan, H.-Y.; Hu, Y.-C. A graphene-based platform for induced pluripotent stem cells culture and differentiation. *Biomaterials* **2012**, *33*, 418–427. [CrossRef] [PubMed]
63. Chen, G.-Y.; Chen, C.-L.; Tuan, H.-Y.; Yuan, P.-X.; Li, K.-C.; Yang, H.-J.; Hu, Y.-C. Graphene Oxide Triggers Toll-Like Receptors/Autophagy Responses In Vitro and Inhibits Tumor Growth In Vivo. *Adv. Healthc. Mater.* **2014**, *3*, 1486–1495. [CrossRef] [PubMed]
64. Hanley, C.; Layne, J.; Punnoose, A.; Reddy, K.M.; Coombs, I.; Coombs, A.; Feris, K.; Wingett, D. Preferential killing of cancer cells and activated human T cells using ZnO nanoparticles. *Nanotechnology* **2008**, *19*, 295103. [CrossRef] [PubMed]
65. Wang, H.; Wingett, D.; Engelhard, M.H.; Feris, K.; Reddy, K.M.; Turner, P.; Layne, J.; Hanley, C.; Bell, J.; Tenne, D.; Wang, C.; Punnoose, A. Fluorescent dye encapsulated ZnO particles with cell-specific toxicity for potential use in biomedical applications. *J. Mater. Sci. Mater. Med.* **2009**, *20*, 11–22. [CrossRef] [PubMed]
66. Colilla, M.; González, B.; Vallet-Regí, M. Mesoporous silicananoparticles for the design of smart delivery nanodevices. *Biomater. Sci.* **2013**, *1*, 114–134. [CrossRef]
67. Baeza, A.; Colilla, M.; Vallet-Regí, M. Advances in mesoporous silica nanoparticles for targeted stimuli-responsive drug delivery. *Expert Opin. Drug Deliv.* **2015**, *12*, 319–337. [CrossRef] [PubMed]
68. Martínez-Carmona, M.; Colilla, M.; Vallet-Regí, M. Smart Mesoporous Nanomaterials for Antitumor Therapy. *Nanomaterials* **2015**, *5*, 1906–1937. [CrossRef] [PubMed]
69. Van Oosten, M.; Schäfer, T.; Gazendam, J.A.C.; Ohlsen, K.; Tsompanidou, E.; De Goffau, M.C.; Harmsen, H.J.M.; Crane, L.M.A.; Lim, E.; Francis, K.P.; et al. Real-time in vivo imaging of invasive- and biomaterial-associated bacterial infections using fluorescently labelled vancomycin. *Nat. Commun.* **2013**, *4*. [CrossRef] [PubMed]
70. Israel, O.; Keidar, Z. PET/CT imaging in infectious conditions. *Ann. N. Y. Acad. Sci.* **2011**, *1228*, 150–166. [CrossRef] [PubMed]
71. Sievert, D.M.; Ricks, P.; Edwards, J.R.; Schneider, A.; Patel, J.; Srinivasan, A.; Kallen, A.; Limbago, B.; Fridkin, S. Antimicrobial-Resistant Pathogens Associated with Healthcare-Associated Infections: Summary of Data Reported to the National Healthcare Safety Network at the Centers for Disease Control and Prevention, 2009–2010. *Infect. Control Hosp. Epidemiol.* **2013**, *34*, 1–14. [CrossRef] [PubMed]
72. Høiby, N.; Bjarnsholt, T.; Givskov, M.; Molin, S.; Ciofu, O. Antibiotic resistance of bacterial biofilms. *Int. J. Antimicrob. Agents* **2018**, *35*, 322–332. [CrossRef] [PubMed]
73. Pinto-Alphandary, H.; Andremont, A.; Couvreur, P. Targeted delivery of antibiotics using liposomes and nanoparticles: Research and applications. *Int. J. Antimicrob. Agents* **2000**, *13*, 155–168. [CrossRef]
74. Dinjaski, N.; Suri, S.; Valle, J.; Lehman, S.M.; Lasa, I.; Prieto, M.A.; García, A.J. Near-infrared fluorescence imaging as an alternative to bioluminescent bacteria to monitor biomaterial-associated infections. *Acta Biomater.* **2014**, *10*, 2935–2944. [CrossRef] [PubMed]
75. Manzoor, U.; Siddique, S.; Ahmed, R.; Noreen, Z.; Bokhari, H.; Ahmad, I. Antibacterial, structural and optical characterization of mechano-chemically prepared ZnO nanoparticles. *PLoS ONE* **2016**, *11*, 1–12. [CrossRef] [PubMed]
76. Reddy, K.M.; Feris, K.; Bell, J.; Wingett, D.G.; Hanley, C.; Punnoose, A. Selective toxicity of zinc oxide nanoparticles to prokaryotic and eukaryotic systems. *Appl. Phys. Lett.* **2007**, *90*, 213902–213903. [CrossRef] [PubMed]
77. Wang, L.; Hu, C.; Shao, L. The antimicrobial activity of nanoparticles: Present situation and prospects for the future. *Int. J. Nanomed.* **2017**, *12*, 1227–1249. [CrossRef] [PubMed]

78. Applerot, G.; Lipovsky, A.; Dror, R.; Perkas, N.; Nitzan, Y.; Lubart, R.; Gedanken, A. Enhanced Antibacterial Activity of Nanocrystalline ZnO Due to Increased ROS-Mediated Cell Injury. *Adv. Funct. Mater.* **2009**, *19*, 842–852. [CrossRef]
79. Raghupathi, K.R.; Koodali, R.T.; Manna, A.C. Size-dependent bacterial growth inhibition and mechanism of antibacterial activity of zinc oxide nanoparticles. *Langmuir* **2011**, *27*, 4020–4028. [CrossRef] [PubMed]
80. Xie, Y.; He, Y.; Irwin, P.L.; Jin, T.; Shi, X. Antibacterial activity and mechanism of action of zinc oxide nanoparticles against Campylobacter jejuni. *Appl. Environ. Microbiol.* **2011**, *77*, 2325–2331. [CrossRef] [PubMed]
81. Jafari, A.R.; Mosavi, T.; Mosavari, N.; Majid, A.; Movahedzade, F.; Tebyaniyan, M.; Kamalzadeh, M.; Dehgan, M.; Jafari, S.; Arastoo, S. Mixed metal oxide nanoparticles inhibit growth of Mycobacterium tuberculosis into THP-1 cells. *Int. J. Mycobacteriol.* **2016**, *5*, S181–S183. [CrossRef] [PubMed]
82. Gupta, J.; Bahadur, D. Visible Light Sensitive Mesoporous Cu-Substituted ZnO Nanoassembly for Enhanced Photocatalysis, Bacterial Inhibition, and Noninvasive Tumor Regression. *ACS Sustain. Chem. Eng.* **2017**, *5*, 8702–8709. [CrossRef]
83. Kadiyala, U.; Tulari-Emre, E.S.; Bahng, J.H.; Kotov, N.A.; VanEpps, J.S. Unexpected insights into antibacterial activity of zinc oxide nanoparticles against methicillin resistant *Staphylococcus aureus* (MRSA). *Nanoscale* **2018**, 4927–4939. [CrossRef] [PubMed]
84. Sehmi, S.K.; Noimark, S.; Bear, J.C.; Peveler, W.J.; Bovis, M.; Allan, E.; MacRobert, A.J.; Parkin, I.P. Lethal photosensitisation of Staphylococcus aureus and Escherichia coli using crystal violet and zinc oxide-encapsulated polyurethane. *J. Mater. Chem. B* **2015**, *3*, 6490–6500. [CrossRef]
85. Chen, H.; Zhang, M.; Li, B.; Chen, D.; Dong, X.; Wang, Y.; Gu, Y. Versatile antimicrobial peptide-based ZnO quantum dots for invivo bacteria diagnosis and treatment with high specificity. *Biomaterials* **2015**, *53*, 532–544. [CrossRef] [PubMed]
86. Aponiene, K.; Serevičius, T.; Luksiene, Z.; Juršėnas, S. Inactivation of bacterial biofilms using visible-light-activated unmodified ZnO nanorods. *Nanotechnology* **2017**, *28*, 365701. [CrossRef] [PubMed]
87. Dwivedi, S.; Wahab, R.; Khan, F.; Mishra, Y.K.; Musarrat, J.; Al-Khedhairy, A.A. Reactive oxygen species mediated bacterial biofilm inhibition via zinc oxide nanoparticles and their statistical determination. *PLoS ONE* **2014**, *9*, 1–9. [CrossRef] [PubMed]
88. Lee, J.H.; Kim, Y.G.; Cho, M.H.; Lee, J. ZnO nanoparticles inhibit Pseudomonas aeruginosa biofilm formation and virulence factor production. *Microbiol. Res.* **2014**, *169*, 888–896. [CrossRef] [PubMed]
89. Shakerimoghaddam, A.; Ghaemi, E.A.; Jamalli, A. Zinc oxide nanoparticle reduced biofilm formation and antigen 43 expressions in uropathogenic Escherichia coli. *Iran. J. Basic Med. Sci.* **2017**, *20*, 451–456. [CrossRef] [PubMed]
90. Sarwar, S.; Chakraborti, S.; Bera, S.; Sheikh, I.A.; Hoque, K.M.; Chakrabarti, P. The antimicrobial activity of ZnO nanoparticles against Vibrio cholerae: Variation in response depends on biotype. *Nanomed. Nanotechnol. Biol. Med.* **2016**, *12*, 1499–1509. [CrossRef] [PubMed]
91. Schey, K.L.; Luther, J.M.; Rose, K.L. Zinc Oxide Nanoparticle Suspensions and Layer-By-Layer Coatings Inhibit Staphylococcal Growth. *Nanomedicine* **2016**, *12*, 1–21. [CrossRef]
92. Alves, M.M.; Bouchami, O.; Tavares, A.; Córdoba, L.; Santos, C.F.; Miragaia, M.; De Fátima Montemor, M. New Insights into Antibiofilm Effect of a Nanosized ZnO Coating against the Pathogenic Methicillin Resistant Staphylococcus aureus. *ACS Appl. Mater. Interfaces* **2017**, *9*, 28157–28167. [CrossRef] [PubMed]
93. Liao, H.; Miao, X.; Ye, J.; Wu, T.; Deng, Z.; Li, C.; Jia, J.; Cheng, X.; Wang, X. Falling Leaves Inspired ZnO Nanorods-Nanoslices Hierarchical Structure for Implant Surface Modification with Two Stage Releasing Features. *ACS Appl. Mater. Interfaces* **2017**, *9*, 13009–13015. [CrossRef] [PubMed]
94. Hsueh, Y.H.; Ke, W.J.; Hsieh, C. Te; Lin, K.S.; Tzou, D.Y.; Chiang, C.L. ZnO nanoparticles affect bacillus subtilis cell growth and biofilm formation. *PLoS ONE* **2015**, *10*, 1–23. [CrossRef] [PubMed]
95. Ghasemi, F.; Jalal, R. Antimicrobial action of zinc oxide nanoparticles in combination with ciprofloxacin and ceftazidime against multidrug-resistant Acinetobacter baumannii. *J. Glob. Antimicrob. Resist.* **2016**, *6*, 118–122. [CrossRef] [PubMed]
96. Spadaro, D.; Garibaldi, A.; Gullino, M.L. Control of Penicillium expansum and Botrytis cinerea on apple combining a biocontrol agent with hot water dipping and acibenzolar-S-methyl, baking soda, or ethanol application. *Postharv. Biol. Technol.* **2004**, *33*, 141–151. [CrossRef]

97. He, L.; Liu, Y.; Mustapha, A.; Lin, M. Antifungal activity of zinc oxide nanoparticles against Botrytis cinerea and Penicillium expansum. *Microbiol. Res.* **2011**, *166*, 207–215. [CrossRef] [PubMed]
98. Surendra, T.V.; Roopan, S.M.; Al-Dhabi, N.A.; Arasu, M.V.; Sarkar, G.; Suthindhiran, K. Vegetable Peel Waste for the Production of ZnO Nanoparticles and its Toxicological Efficiency, Antifungal, Hemolytic, and Antibacterial Activities. *Nanoscale Res. Lett.* **2016**, *11*. [CrossRef] [PubMed]
99. Gunalan, S.; Sivaraj, R.; Rajendran, V. Green synthesized ZnO nanoparticles against bacterial and fungal pathogens. *Prog. Nat. Sci. Mater. Int.* **2012**, *22*, 693–700. [CrossRef]
100. Arciniegas-Grijalba, P.A.; Patiño-Portela, M.C.; Mosquera-Sánchez, L.P.; Guerrero-Vargas, J.A.; Rodríguez-Páez, J.E. ZnO nanoparticles (ZnO-NPs) and their antifungal activity against coffee fungus Erythricium salmonicolor. *Appl. Nanosci.* **2017**, *7*, 225–241. [CrossRef]
101. Sierra-Fernandez, A.; De La Rosa-García, S.C.; Gomez-Villalba, L.S.; Gómez-Cornelio, S.; Rabanal, M.E.; Fort, R.; Quintana, P. Synthesis, Photocatalytic, and Antifungal Properties of MgO, ZnO and Zn/Mg Oxide Nanoparticles for the Protection of Calcareous Stone Heritage. *ACS Appl. Mater. Interfaces* **2017**, *9*, 24873–24886. [CrossRef] [PubMed]
102. Jasim, N.O. Antifungal Activity of Zinc Oxide Nanoparticles on Aspergillus Fumigatus Fungus & Candida Albicans Yeast. *Citeseer* **2015**, *5*, 23–28.
103. Hong, R.Y.; Li, J.H.; Chen, L.L.; Liu, D.Q.; Li, H.Z.; Zheng, Y.; Ding, J. Synthesis, surface modification and photocatalytic property of ZnO nanoparticles. *Powder Technol.* **2009**, *189*, 426–432. [CrossRef]
104. Palmquist, D.L.; Lock, A.L.; Shingfield, K.J.; Bauman, D.E. Biosynthesis of Conjugated Linoleic Acid in Ruminants and Humans. In *Advances in Food and Nutrition Research*; Academic Press: Cambridge, MA, USA, 2005; Volume 50, pp. 179–217.
105. Barad, S.; Roudbary, M.; Omran, N.A.; Daryasari, P.M. Preparation and characterization of ZnO nanoparticles coated by chitosan-linoleic acid; fungal growth and biofilm assay. *Bratisl. Med. J.-Bratisl. Lek. List.* **2017**, *118*, 169–174. [CrossRef]
106. Decelis, S.; Sardella, D.; Triganza, T.; Brincat, J.-P.; Gatt, R.; Valdramidis, V.P. Assessing the anti-fungal efficiency of filters coated with zinc oxide nanoparticles. *R. Soc. Open Sci.* **2017**, *4*, 1–9. [CrossRef] [PubMed]
107. Esteban-Tejeda, L.; Prado, C.; Cabal, B.; Sanz, J.; Torrecillas, R.; Moya, J.S. Antibacterial and antifungal activity of ZnO containing glasses. *PLoS ONE* **2015**, *10*. [CrossRef] [PubMed]
108. National Diabetes Statistics Report. 2017. Available online: https://www.cdc.gov/diabetes/data/statistics-report/index.html (accessed on 23 April 2018).
109. Abel, J.J. Crystalline Insulin. *Proc. Natl. Acad. Sci. USA* **1926**, *12*, 132–136. [CrossRef] [PubMed]
110. Ranasinghe, P.; Pigera, S.; Galappatthy, P.; Katulanda, P.; Constantine, G.R. Zinc and diabetes mellitus: Understanding molecular mechanisms and clinical implications. *DARU J. Pharm. Sci.* **2015**, *23*, 1–13. [CrossRef] [PubMed]
111. Umrani, R.D.; Paknikar, K.M. Zinc oxide nanoparticles show antidiabetic activity in streptozotocin-induced Type 1 and 2 diabetic rats. *Nanomedicine* **2014**, *9*, 89–104. [CrossRef] [PubMed]
112. Nazarizadeh, A.; Asri-Rezaie, S. Comparative Study of Antidiabetic Activity and Oxidative Stress Induced by Zinc Oxide Nanoparticles and Zinc Sulfate in Diabetic Rats. *AAPS PharmSciTech* **2016**, *17*, 834–843. [CrossRef] [PubMed]
113. Wahba, N.S.; Shaban, S.F.; Kattaia, A.A.A.; Kandeel, S.A. Efficacy of zinc oxide nanoparticles in attenuating pancreatic damage in a rat model of streptozotocin-induced diabetes. *Ultrastruct. Pathol.* **2016**, *40*, 358–373. [CrossRef] [PubMed]
114. Shanker, K.; Naradala, J.; Mohan, G.K.; Kumar, G.S.; Pravallika, P.L. A sub-acute oral toxicity analysis and comparative in vivo anti-diabetic activity of zinc oxide, cerium oxide, silver nanoparticles, and Momordica charantia in streptozotocin-induced diabetic Wistar rats. *RSC Adv.* **2017**, *7*, 37158–37167. [CrossRef]
115. El-Gharbawy, R.M.; Emara, A.M.; Abu-Risha, S.E.S. Zinc oxide nanoparticles and a standard antidiabetic drug restore the function and structure of beta cells in Type-2 diabetes. *Biomed. Pharmacother.* **2016**, *84*, 810–820. [CrossRef] [PubMed]
116. Kitture, R.; Chordiya, K.; Gaware, S.; Ghosh, S.; More, P.A.; Kulkarni, P.; Chopade, B.A.; Kale, S.N. ZnO Nanoparticles-Red Sandalwood Conjugate: A Promising Anti-Diabetic Agent. *J. Nanosci. Nanotechnol.* **2015**, *15*, 4046–4051. [CrossRef] [PubMed]
117. Gupta, M.; Mahajan, V.K.; Mehta, K.S.; Chauhan, P.S. Zinc therapy in dermatology: A review. *Dermatol. Res. Pract.* **2014**, *2014*. [CrossRef] [PubMed]

118. Ilves, M.; Palomaki, J.; Vippola, M.; Lehto, M.; Savolainen, K.; Savinko, T.; Alenius, H. Topically applied ZnO nanoparticles suppress allergen induced skin inflammation but induce vigorous IgE production in the atopic dermatitis mouse model. *Part. Fibre Toxicol.* **2014**, *11*, 1–12. [CrossRef] [PubMed]
119. Nagajyothi, P.C.; Cha, S.J.; Yang, I.J.; Sreekanth, T.V.M.; Kim, K.J.; Shin, H.M. Antioxidant and anti-inflammatory activities of zinc oxide nanoparticles synthesized using Polygala tenuifolia root extract. *J. Photochem. Photobiol. B Biol.* **2015**, *146*, 10–17. [CrossRef] [PubMed]
120. Li, J.; Chen, H.; Wang, B.; Cai, C.; Yang, X.; Chai, Z.; Feng, W. ZnO nanoparticles act as supportive therapy in DSS-induced ulcerative colitis in mice by maintaining gut homeostasis and activating Nrf2 signaling. *Sci. Rep.* **2017**, *7*, 1–11. [CrossRef] [PubMed]
121. Singh, B.N.; Rawat, A.K.S.; Khan, W.; Naqvi, A.H.; Singh, B.R. Biosynthesis of stable antioxidant ZnO nanoparticles by Pseudomonas aeruginosa Rhamnolipids. *PLoS ONE* **2014**, *9*. [CrossRef] [PubMed]
122. Agren, M.S. Studies on zinc in wound-healing. *Acta Derm. Venereol.* **1990**, 1–36.
123. Lansdown, A.B.G.; Mirastschijski, U.; Stubbs, N.; Scanlon, E.; Ågren, M.S. Zinc in wound healing: Theoretical, experimental, and clinical aspects. *Wound Repair Regen.* **2007**, *15*, 2–16. [CrossRef] [PubMed]
124. Kogan, S.; Sood, A.; Granick, M.S. Zinc and Wound Healing: A Review of Zinc Physiology and Clinical Applications. *WOUNDS-A Compend. Clin. Res. Pract.* **2017**, *29*, 102–106.
125. Sudheesh Kumar, P.T.; Lakshmanan, V.K.; Anilkumar, T.V.; Ramya, C.; Reshmi, P.; Unnikrishnan, A.G.; Nair, S.V.; Jayakumar, R. Flexible and microporous chitosan hydrogel/nano ZnO composite bandages for wound dressing: In vitro and in vivo evaluation. *ACS Appl. Mater. Interfaces* **2012**, *4*, 2618–2629. [CrossRef] [PubMed]
126. Raguvaran, R.; Manuja, B.K.; Chopra, M.; Thakur, R.; Anand, T.; Kalia, A.; Manuja, A. Sodium alginate and gum acacia hydrogels of ZnO nanoparticles show wound healing effect on fibroblast cells. *Int. J. Biol. Macromol.* **2017**, *96*, 185–191. [CrossRef] [PubMed]
127. Croisier, F.; Jérôme, C. Chitosan-based biomaterials for tissue engineering. *Eur. Polym. J.* **2013**, *49*, 780–792. [CrossRef]
128. Karahaliloglu, Z.; Kilicay, E.; Denkbas, E.B. Antibacterial chitosan/silk sericin 3D porous scaffolds as a wound dressing material. *Artif. Cells Nanomed. Biotechnol.* **2017**, *45*, 1172–1185. [CrossRef] [PubMed]
129. Cahú, T.B.; Silva, R.A.; Silva, R.P.F.; Silva, M.M.; Arruda, I.R.S.; Silva, J.F.; Costa, R.M.P.B.; Santos, S.D.; Nader, H.B.; Bezerra, R.S. Evaluation of Chitosan-Based Films Containing Gelatin, Chondroitin 4-Sulfate and ZnO for Wound Healing. *Appl. Biochem. Biotechnol.* **2017**, *183*, 765–777. [CrossRef] [PubMed]

© 2018 by the authors. Licensee MDPI, Basel, Switzerland. This article is an open access article distributed under the terms and conditions of the Creative Commons Attribution (CC BY) license (http://creativecommons.org/licenses/by/4.0/).

Article

A Microwave-Assisted Synthesis of Zinc Oxide Nanocrystals Finely Tuned for Biological Applications

Nadia Garino [1,2], Tania Limongi [1], Bianca Dumontel [1], Marta Canta [1], Luisa Racca [1], Marco Laurenti [1], Micaela Castellino [1], Alberto Casu [3], Andrea Falqui [3] and Valentina Cauda [1,*]

[1] Department of Applied Science and Technology, Politecnico di Torino, Corso Duca degli Abruzzi 24, 10129 Turin, Italy; nadia.garino@polito.it (N.G.); tania.limongi@polito.it (T.L.); bianca.dumontel@polito.it (B.D.); marta.canta@polito.it (M.C.); luisa.racca@polito.it (L.R.); marco.laurenti@polito.it (M.L.); micaela.castellino@polito.it (M.C.)
[2] Istituto Italiano di Tecnologia, Center for Sustainable Future Technologies, Via Livorno 60, 10144 Torino, Italy
[3] King Abdullah University of Science and Technology (KAUST), Biological and Engineering (BESE) Division, NABLA Lab, Thuwal 23955, Saudi Arabia; alberto.casu@kaust.edu.sa (A.C.); andrea.falqui@kaust.edu.sa (A.F.)
* Correspondence: valentina.cauda@polito.it; Tel.: +39-011-090-7389

Received: 14 January 2019; Accepted: 30 January 2019; Published: 6 February 2019

Abstract: Herein we report a novel, easy, fast and reliable microwave-assisted synthesis procedure for the preparation of colloidal zinc oxide nanocrystals (ZnO NCs) optimized for biological applications. ZnO NCs are also prepared by a conventional solvo-thermal approach and the properties of the two families of NCs are compared and discussed. All of the NCs are fully characterized in terms of morphological analysis, crystalline structure, chemical composition and optical properties, both as pristine nanomaterials or after amino-propyl group functionalization. Compared to the conventional approach, the novel microwave-derived ZnO NCs demonstrate outstanding colloidal stability in ethanol and water with long shelf-life. Furthermore, together with their more uniform size, shape and chemical surface properties, this long-term colloidal stability also contributes to the highly reproducible data in terms of biocompatibility. Actually, a significantly different biological behavior of the microwave-synthesized ZnO NCs is reported with respect to NCs prepared by the conventional synthesis procedure. In particular, consistent cytotoxicity and highly reproducible cell uptake toward KB cancer cells are measured with the use of microwave-synthesized ZnO NCs, in contrast to the non-reproducible and scattered data obtained with the conventionally-synthesized ones. Thus, we demonstrate how the synthetic route and, as a consequence, the control over all the nanomaterial properties are prominent points to be considered when dealing with the biological world for the achievement of reproducible and reliable results, and how the use of commercially-available and under-characterized nanomaterials should be discouraged in this view.

Keywords: zinc oxide; microwave solvothermal synthesis; hydrodynamic size; surface chemistry; nanocrystals; cell cytotoxicity

---

1. Introduction

In the last decade metal oxide semiconducting nanoparticles (NPs) have been receiving great interest in the field of biological applications due to their intriguing optical properties, low toxicity, good biocompatibility and their low cost [1]. Among them, zinc oxide (ZnO) nanoparticles have shown to be particularly promising due to their peculiar chemical and physical properties, which can be specifically tailored on the basis of the particles' size and shape [2,3]. As an example, the wide bandgap typical of ZnO (about 3.37 eV at room temperature, RT) entails a fluorescence excitation situated in the ultraviolet (UV) region [4], which allows ZnO to be successfully employed for optical cell imaging. More in general, it has been demonstrated how ZnO could be a promising material for therapeutic and diagnostic applications [2], showing high levels of drugs loading and a quite easy control over the following release [5,6]. In this context, the optical, targeting and drug delivery properties of ZnO can be more specifically addressed by the combination of various synthetic procedures (sol-gel, sputtering, hydro-solvothermal, etc.) [7] and ZnO morphologies (nanowires, nanorods, nanobelts, desert roses and spherical nanoparticles) [8,9], together with surface functionalization approaches [2,10]. To this purpose, it is important to observe that crystal density, morphology and defects could be critical factors in determining the material properties, and as a consequence, the final applications. Although ZnO nanostructures featuring dimensions lower than 100 nm are nowadays considered the most promising for biomedicine [11–13], the most part of the literature generally neglects the influence of the synthesis processes and parameters on the characteristics of the resulting NPs and, finally, on the reproducibility of their biological response. Actually, this is an essential point for further strengthening the biological application of crystalline nanoparticles. In this regard, it is worth mentioning that many articles deal with the use of commercial particles (with the limit about their morphologies and size distribution) or did not report any specific detail about important synthesis aspects like the synthesis precursors, the surface chemistry, and sometimes the hydrodynamic size and z-potential [14,15].

In this work, we report a novel synthetic approach of ZnO nanocrystals (NCs) based on a microwave-assisted solvothermal process that allows us to reach a greater control over the morphology and dimensional dispersion of the NPs. At the same time this new method guarantees a higher reproducibility level of experimental data with respect to those obtained by using ZnO NCs synthesized with a more conventional wet approach. The microwave-assisted synthesis also presents additional advantages, like shorter reaction time and lower energy consumption. Actually, one of the main characteristics of this technique is to guarantee a uniform heating of the precursors and to present outstanding reaction rates, moving from several hours to a few minutes [16–18]. Furthermore, this synthetic method assures high reliability and high reaction yields. This explains the growing popularity and diffusion of microwave-assisted synthetic approaches over a wide range of different nanomaterials synthesis [19,20], but to our knowledge very few reports are dedicated to the preparation of ZnO-based nanomaterials [21–24]. Indeed, thanks to this synthetic route we obtain a simultaneous nucleation of nanocrystals that leads to a uniform and reproducible dimensional dispersion of ZnO (about 20 nm in diameter), with outstanding colloidal stability both in ethanol and in water. In this way, we demonstrate how the synthetic route should be an important factor to be considered for the achievement of reproducible and reliable results, also when evaluating the cell viability when in contact with these nanocrystals.

2. Materials and Methods

2.1. Synthesis, Functionalization and Labelling of ZnO Nanocrystals.

All the chemicals were used as purchased without further purification. ZnO nanocrystals were synthesized through two different synthetic routes: a traditional solvothermal way (sample named ZnO-st) and a microwave-assisted synthesis (sample named ZnO-mw). As zinc precursors we chose zinc acetate di-hydrate ($Zn(CH_3COO)_2 \cdot 2H_2O$ Puriss. p.a., ACS Reagent, $\geq 99.0\%$ Fluka) and a hydroxide as mineralizing agent (Sigma-Aldrich) both dissolved in methanol (Reag. Ph Eur Grade VWR Chemicals). The reaction path, in both cases, is based on the hydrolysis of the zinc precursor due to the presence of the hydroxide, as shown in the following reactions scheme:

$$KOH \rightarrow K^+ (aq) + OH^- (aq) \tag{1}$$

$$Zn(Ac)_2 \rightarrow Zn^{2+} (aq) + 2Ac^- (aq) \tag{2}$$

$$Zn^{2+} (aq) + 2OH^- (aq) \rightarrow Zn(OH)_2 (aq) \tag{3}$$

$$Zn(OH)_2 \text{ (aq)} + 2H_2O \rightarrow Zn(OH)_4^{2-} + 2H^+ \text{ (aq)} \qquad (4)$$

$$Zn(OH)_4^{2-} + 2H^+ \text{ (aq)} \rightarrow ZnO \text{ (s)} + 3H_2O \qquad (5)$$

The microwave-assisted synthesis was carried out as follows. A solution containing the zinc precursor in methanol (0.09 M, 60 mL) was prepared and stirred directly in the Teflon reactor vessel. In order to better initiate the zinc oxide nucleation, 480 μL of double-distilled water was added and then the potassium hydroxide solution (0.2 M, 35 mL) (KOH ≥ 85% pellets, Sigma-Aldrich) was mixed together in a 270 mL Teflon reactor vessel, equipped with pressure and temperature probes, connected with the microwave furnace (Milestone START-Synth, Milestone Inc, Shelton, Connecticut). The resulting solution was put into a microwave oven for 30 min at 60 °C. After the completion of the reaction, the solution was cooled down to room temperature and followed by two washing steps to change the reaction solvent and to remove any unreacted compound. To do that, the colloidal solution was collected and centrifuged for 10 min at 3500 g (Mega Star 600R, VWR), the supernatant was then removed, and the precipitate was dispersed and washed twice in 15 mL of ethanol (Sigma-Aldrich, 99%). The as-obtained ZnO NCs pellet was suspended through sonication (LABSONIC LBS2, FALC Instruments SRL) in fresh ethanol to give the final colloidal suspension.

The traditional solvothermal synthesis process was carried out as already reported [25] with the same zinc precursor at the same concentrations in a round-bottom glass flask (100 mL). In detail, zinc acetate di-hydrate (0.09 M) was directly dissolved in the reaction flask with methanol (42 mL) and heated under continuous stirring in reflux conditions since the temperature of 60 °C was reached and the double-distilled water was added (318 μL). The methanol solution of sodium hydroxide (0.31 M, 23 mL) (NaOH BioXtra, ≥98% acidimetric, pellets anhydrous, Sigma-Aldrich) was then added dropwise to the zinc acetate solution (in about 20 min). The reaction conditions were maintained for 2.5 h and after this time the as obtained suspension was cooled to RT. ZnO NCs (named ZnO-st) were collected and washed with fresh ethanol (Sigma-Aldrich, 99%) as previously reported for the microwave-assisted synthesis. Reaction yields were evaluated for both the synthetic procedures by weighing the dried NCs from a known volume of the obtained colloidal solutions.

Both the typologies of as synthesized ZnO NCs were functionalized in order to proceed with the in vitro cell culture studies, according to a previously reported method [25,26]. In particular, the ZnO NCs surface was decorated with the amino-propyl group (ZnO-NH_2 NCs). Approximately 50 mg of ZnO NCs, dispersed in 20 mL of ethanol (Sigma-Aldrich 99%), were heated to 80 °C in a 25 mL round glass flask under continuous stirring and nitrogen gas flow. A 10 mol% ratio of 3-aminopropyltrimethoxysilane ($H_2N(CH_2)_3Si(OCH_3)_3$ APTMS 97%, Sigma Aldrich, 10 μL), with respect to total ZnO amount, was added to the NCs suspension and the reaction was carried out for 6 h. The excess of unreacted APTMS was then removed by washing twice the ZnO NCs with fresh ethanol, separating them from the reaction medium by centrifugation (10 min, 10,000 g).

Only for the internalization of nanocrystals into cancer cells, the ZnO-NH_2 NCs were coupled with ATTO633-NHS ester dyes (Thermofischer), by adding 2 μg of dye each mg of NCs in ethanol suspension. The as obtained solution was dark-stirred overnight and then washed twice by centrifuging (10 min, 10,000 g) and resuspending the pellet in fresh ethanol to remove unbounded dye molecules [25]. To minimize the effects of particles aggregation and sedimentation under biological tests, the suspension of dye-labelled nanocrystals was always freshly prepared and shortly sonicated (10 min) before each experiment.

2.2. Characterization

The crystalline structure of the prepared materials was investigated by XRD (X-ray diffraction). A Panalytical X'Pert diffractometer in θ–2θ Bragg-Brentano configuration equipped with a source of radiation Cu-Kα (λ = 1.54 Å, 40 kV and 30 mA) was employed. The samples were prepared depositing the colloidal solution drop by drop on a silicon wafer. The XRD analysis was carried out at room temperature in the 2θ range 20°–65° with a step size of 0.02° (2θ), and an acquisition time of 100 s per step.

Further information on the chemical composition and surface of the produced materials was provided by X-ray photoelectron spectroscopy (XPS) analysis. A PHI 5000 Versaprobe Scanning X-ray photoelectron spectrometer (monochromatic Al K-alpha X-ray source with 1486.6 eV energy) was used to investigate the material chemical composition. A spot size of 100 μm was used in order to collect the photoelectron signal for both the high resolution (HR) and the survey spectra.

Field emission scanning electron microscopy (FESEM, Merlin, ZEISS, Jena, Germany) was used to evaluate the overall quality and the morphology of the different materials. The samples were prepared by depositing a drop of a properly diluted NCs solution on top of silicon wafer.

Both conventional and high-resolution transmission electron microscopy (CTEM and HRTEM) were used to characterize the morphological and structural features of the different materials. CTEM was performed by using a FEI Tecnai Spirit microscope working at an acceleration voltage of 120 kV, equipped with a Twin objective lens, a LaB_6 thermionic electron source and a Gatan Orius CCD camera. HRTEM was performed by using a FEI Titan ST microscope working at an acceleration voltage of 300 kV, equipped with a S-Twin objective lens, an ultra-bright field emission electron source (X-FEG) and a Gatan 2 k × 2 k CCD camera.

The hydrodynamic size of the nanocrystals in ethanol and water was determined using the dynamic light scattering (DLS) technique with a Zetasizer Nano ZS90 (Malvern Instruments, Worcestershire, UK). All the measurements were performed at room temperature, at a concentration of 100 μg/mL sonicating each sample for 10 min before the acquisition.

UV−visible spectra were recorded in absorbance with a Multiskan GO microplate UV−Vis spectrophotometer (Thermofisher Scientific) using the ZnO NCs in ethanolic suspension at a concentration of 0.5 mg/mL. All the spectra were background subtracted.

Fluorescence excitation and emission spectra were recorded by a Perkin Elmer LS55 fluorescence spectrometer (PerkinElmer Inc. Waltham, Massachusetts, MA, USA) using quartz cuvettes of 1 cm optical path and containing the ZnO NCs in ethanolic suspension at a concentration of 0.5 mg/mL. For the emission spectra, the fluorescence excitation wavelength used was 380 nm and, for the excitation spectra, the fluorescence emission wavelength was 500 nm. Scans were acquired with 2.5 slits opening and at a scan rate of 300 nm/min.

2.3. In Vitro Mammalian Cell Culture Biological Test

Cell culture and reagents. KB cell line (ATCC® CCL17TM) was purchased by the American Type Culture Collection. Cells were grown in minimal essential Eagle's medium (EMEM, Sigma) supplemented with 10% heath inactivated fetal bovine serum (FBS, Sigma), 100 units/mL penicillin and 100 μg/mL streptomycin (Sigma) and maintained at 37 °C under a 5% CO_2 atmosphere. Cells were periodically tested for mycoplasma infection.

All the NCs solutions were freshly prepared from 1 mg/mL ethanol stock solutions. The ZnO NCs were bath sonicated at 40 kHz, 100% power for 10 min before being added to the culture medium and then immediately used for treating cells.

Cytotoxicity tests. 1.5×10^3 cells/well were seeded onto 96-well plastic culture plates (Corning® 96 Well TC-Treated Microplates) and incubated at 37 °C, 5% CO_2. After 24 h the cell medium was replaced with fresh medium containing ZnO-st-NH_2 or ZnO-mw-NH_2 at different concentrations (10, 15, 20, 25 μg/mL). After 24 h of incubation, cell proliferation was assessed by measuring cell metabolic activity through the WST-1 cell proliferation assay. 10 μL of the WST-1 reagent (Roche) were added to each well and after 2 h incubation in the dark at 37 °C, 5% CO_2, the formazan absorbance was measured at 490 nm by the Multiskan GO microplate spectrophotometer (Thermofisher Scientific) using a 620 nm reference. Control values (without treatments) were set at 100% viable and all values were expressed as a percentage of the control. The inhibitory concentration 50% (IC50) of the ZnO-mw-NH_2 was determined online using the IC50 calculator tool of the AAT Bioquest webpage [27].

Cell internalization. NCs internalization in KB cells was measured using a Guava Easycyte 6-2L flow cytometer (Merck Millipore). 1×10^5 cells/well were seeded onto a 6-well plate (Corning®

Costar® TC-Treated) with complete cell culture medium 24 h prior the assay. Then, cells were treated with 10 µg/mL of ZnO-st-NH$_2$ or ZnO-mw-NH$_2$ labeled with ATTO633-NHS ester dye (Thermofisher). A control well, containing the untreated cells, was instead filled with fresh medium without NCs. After 24 h incubation, cells were rinsed twice with phosphate buffered saline (PBS), trypsinized and centrifuged at 130 g for 5 min. Cell pellets were re-suspended in 1 mL PBS and immediately analysed with the flow cytometer; 10,000 gated events were considered for the analysis excluding cellular debris, characterized by low FSC (forward scatter) and SSC (Side scatter). Results were reported as the percentage of fluorescence positive events, characterized by a shift in fluorescence intensity compared to untreated cells. This evaluation was performed by Guava InCyte Software (Merck Millipore, Darmstadt, Germany). Independent experiments were performed 3 times. The different NCs internalization was compared using a t test and when equal variance test failed, Welch's test was assumed instead of Student's t test.

Statistical analysis. All experiments were done at least in triplicate and the results were presented as mean ± SEM (standard error of mean). The experimental data were analysed using Sigmaplot software version 14, demo version (Systat Software Inc., San Jose, California, CA, USA). t test and One Way Analysis of Variance were executed, where equal variances were not assumed, the results of Welch's test were presented. When the normality test failed, Mann–Whitney rank sum test was run. Differences were considered significant at p value < 0.05.

3. Results and Discussion

Both synthetic approaches allowed us to obtain milky-coloured colloidal suspensions of ZnO NCs in ethanol. The yields comparison shows that larger yields (i.e., five times higher) are obtained for the microwave assisted procedure with respect to the traditional wet route. The functionalization with amino-propyl groups provides on the one hand a positively charged surface, useful to improve the colloidal stability of the NCs in solution. On the other hand, the amine groups are an ideal anchoring site for dye labelling, commonly used for fluorescence microscopy experiments, as well as for the further functionalization with polymers, lipids or biomolecules useful to enhance the interaction with cells and biological fluids [25,26].

3.1. Morphological and Structural Characterization of ZnO Nanocrystals (NCs)

XRD is performed in order to obtain information about phase identification and quantification, percentage of crystallinity, crystallite size and unit cell size. The XRD patterns of pristine ZnO NCs from both synthetic routes are reported in Figure 1. The comparison of the diffractograms with the standard XRD pattern of ZnO (JCPDS card n. 36-1451) confirms the crystalline structure of the particles. In particular, the peaks at 31.9°, 34.4°, 36.4°, 47.6°, 56.7°, 62.9° are indexed to (100), (002), (101), (102), (110) and (103) planes, respectively, which corresponds to the Miller index of typical hexagonal wurtzite structure. The strongest reflection (101) of each XRD pattern was considered to estimate the average crystallites size with the Debye–Scherrer equation, obtaining a mean diameter of 10.5 nm for ZnO-st NCs and 15.5 nm for ZnO-mw NCs, respectively.

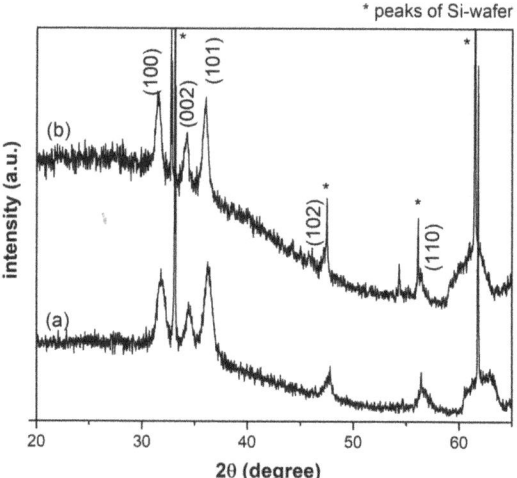

Figure 1. X-ray diffractogram of the pristine ZnO nanocrystals (NCs) obtained via (**a**) conventional (ZnO-st NCs) and (**b**) microwave-assisted synthesis (ZnO-mw NCs).

In order to evaluate the synthesis' repeatability, all the samples are actually characterized by comparing different batches obtained by the two synthetic procedures. FESEM images (here reported in Figure 2) show ZnO spherical morphology for the microwave synthetic route, whereas different shapes, even some elongated ones, are observed for the conventionally-synthesized NCs. By comparing dimensional measurements, we note that the NPs obtained via microwave-assisted synthesis highlight a narrower and reproducible range of particle size distribution, with respect to the traditional one. Indeed, the size distribution of ZnO-mw NCs presents an average size of 20 nm (±5 nm), while for ZnO-st NCs the dimensional range varies between 6 and 20 nm.

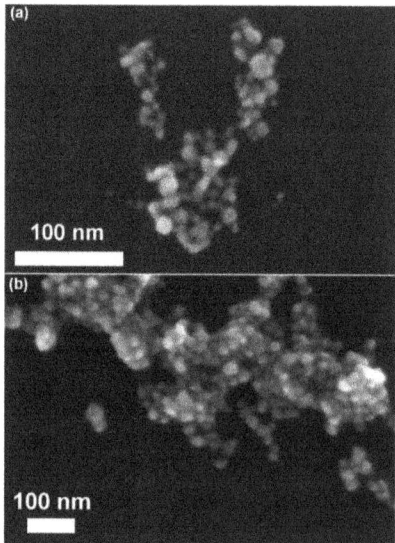

Figure 2. Field emission scanning electron microscopy (FESEM) images of pristine ZnO NCs obtained via (**a**) conventional synthesis and (**b**) microwave-assisted route.

Transmission electron microscopy is also carried out to investigate the structure and crystallinity of ZnO and ZnO-NH$_2$ nanocrystals synthesized with both methods. Thanks to the higher magnification and resolution, it is possible to better highlight the differences between the two as-obtained NC populations. The ZnO-st NCs are reported in Figure 3, where panels (a) and (c) show the CTEM and HRTEM images of the pristine (not functionalized) ZnO-st sample. These particles have a short rod-like shape, whose length goes from 7 to 40 nm and an almost constant width, around 7–8 nm. Besides, HRTEM clearly indicates that these rod-like structures have monocrystalline nature, with no evidence of defects, and with lattice sets' d-spacing and corresponding angular distances expected for the wurtzite structure of ZnO. Panel (b) and (d) of the Figure 3 show what happens to the same ZnO-st particles after functionalization with the amino-propyl groups. Two apparent differences are observed with respect to the non-functionalized crystals: (i) the nanoparticles tend to aggregate, and (ii) the clear presence of an amorphous shell is noticed, surrounding the NCs and clustering them. Again, the HRTEM analysis shows the wurtzite hexagonal structure expected for the ZnO.

Figure 4 displays the ZnO-mw nanocrystals where panels (a) and (c) show the pristine NCs, and panels (b) and (d) those functionalized with the amino-propyl groups. When comparing these images to the two groups prepared by the traditional solvothermal synthesis and displayed in Figure 3, some important differences and analogies have to be highlighted for both. First, the ZnO-mw NCs show a spherical, often faceted, morphology. Their size ranges from 15 to 25 nm, confirming the results obtained by FESEM characterization. HRTEM indicates that both the pristine and the functionalized nanocrystals have the wurtzite hexagonal monocrystalline structure with lattice sets d-spacing and corresponding angular distances expected for the zinc oxide material. Finally, as in the case of the functionalized ZnO-st NCs, no difference in terms of size can be observed between the pristine and amine-functionalized ZnO-mw ones, with the latter displaying an evident propensity to clustering as well as the presence of an amorphous external shell surrounding them.

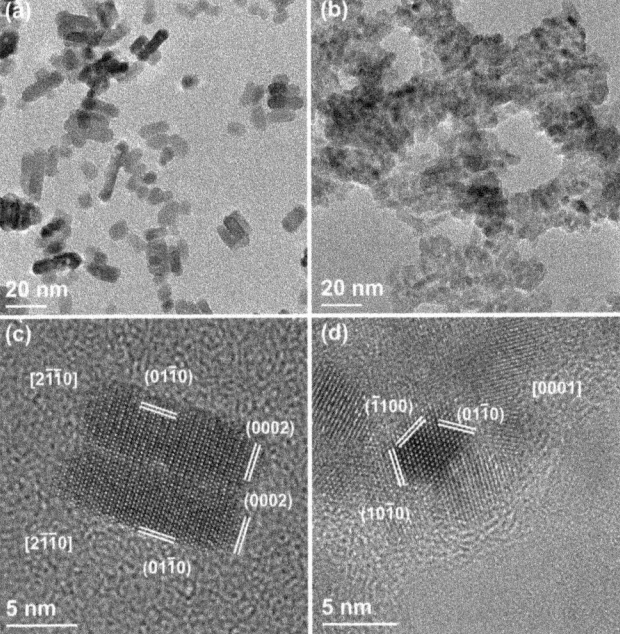

Figure 3. (a,b) Conventional transmission electron microscopy (CTEM) and (c,d) high-resolution transmission electron microscopy (HRTEM) images of (a,c) pristine ZnO-st NCs and (b,d) functionalized ZnO-st-NH$_2$, both obtained via solvothermal synthesis.

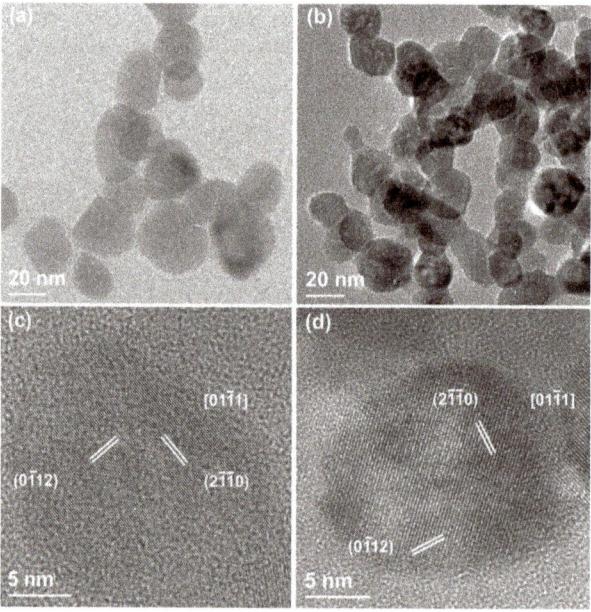

Figure 4. (**a,b**) CTEM and (**c,d**) HRTEM images of (**a,c**) pristine ZnO-mw NCs and (**b,d**) functionalized ZnO-mw-NH$_2$, both obtained via microwave-assisted synthesis.

The colloidal stability of the nanocrystals synthesized by both methods and with or without amine-group functionalization was evaluated by DLS measurements, testing the behavior of three different batches in both ethanol and water and over time. Actually, to show the repeatability (or not) of the proposed synthetic approaches, three different batches are reported in three different color lines (blue, green and red) in Figure 5a,b.

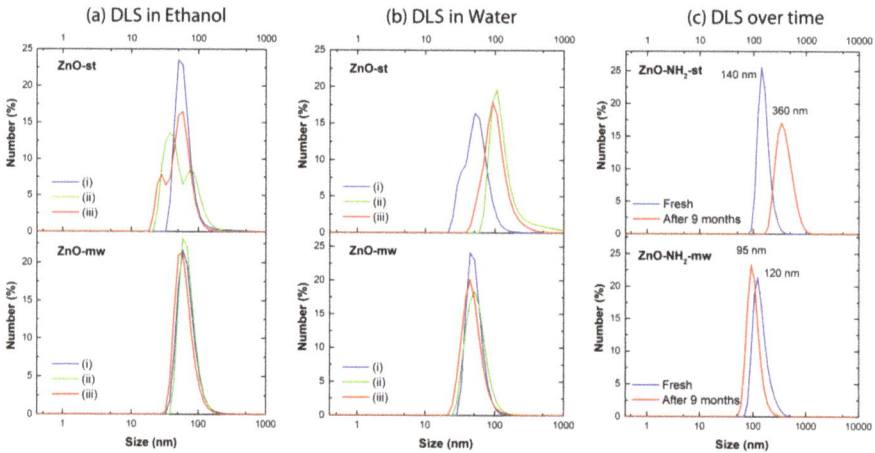

Figure 5. Dynamic light scattering (DLS) measurements in number % of pristine ZnO NCs obtained via traditional synthesis (ZnO-st, top panel) and via microwave route (ZnO-mw, bottom panel) in (**a**) ethanol; (**b**) water; (**c**) of functionalized ZnO particles (ZnO-NH$_2$-mw and ZnO-NH$_2$-st) in ethanol just after the functionalization procedure (solid line) and after nine months of storage (dashed line).

All the pristine nanocrystals obtained via microwave-assisted synthesis (ZnO-mw) result in being well dispersed in both ethanol and water media (Figure 5a,b, bottom panels). These ZnO-mw NCs show narrow size distributions, centered between 50 and 60 nm, and polydispersity indexes (PDI) of less than 0.2, characteristic of monodisperse samples. The nanocrystals synthesized via the traditional solvothermal approach (ZnO-st) present in contrast broader size distributions with higher PDI values (Figure 5a,b, top panels) than those obtained for the ZnO-mw samples, in both ethanol and water media. Furthermore, in some cases, the presence of different size distribution peaks is observed (Figure 5a, top panel). The lower colloidal stability of ZnO-st samples is even more evident in water (Figure 5b, top panel), where the PDI values are considerably higher and the size distributions of some of the tested batches are shifted towards bigger hydrodynamic diameter, indicating an aggregation of the sample (green and red curves in Figure 5b, top panel). A summary of PDI indexes and of mean diameter of number-weighted distributions is reported in Table 1.

The amine-functionalized nanocrystals show a good colloidal stability in ethanol, immediately after functionalization, as reported by the blue lines in Figure 5c (the top panel refers to the ZnO-st NCs, whereas the bottom panel to the ZnO-mw NCs). In the light of the clustering observed by TEM imaging and in order to verify the long-term colloidal stability and the shelf-life of NCs suspension for the biological application, the DLS measurements were also performed on the same batch of ZnO-NH$_2$-st and ZnO-NH$_2$-mw right after the functionalization procedure and after nine months of storage in ethanol. All the samples were subjected to 10 min of ultrasounds before the DLS analyses. The results (summarized in Table 1) indicate a reasonably good stability of ZnO-NH$_2$-mw sample, with a mean hydrodynamic diameter that shift from 120 nm (blue curve) to 96 nm after nine months of storage (red curve, Figure 5c, bottom panel). In contrast, the ZnO-NH$_2$-st NCs present a consistent increase of the mean hydrodynamic diameter (from 140 nm, blue curve, to 360 nm after nine months, red curve in Figure 5c, top panel) indicating an instability and a tendency to aggregation of NCs during the storage. These results also confirm what observed by TEM, despite the very different sample preparation, where the nanocrystals are dried on a copper grid in view of the TEM analysis, thus naturally tending to aggregate, whereas the DLS is performed in solution.

Table 1. Polydispersity indexes (PDI) and average diameter of number-weighted distributions of pristine ZnO NCs obtained via traditional synthesis (ZnO-st NCs) and via microwave route (ZnO-mw NCs) in ethanol and water and of functionalized ZnO particles (ZnO-NH$_2$-mw and ZnO-NH$_2$-st) in ethanol just after the functionalization procedure and after nine months of storage.

		Ethanol			Water				Ethanol	
Sample		Av. diameter (nm)	PDI		Av. diameter (nm)	PDI	Sample		Av. diameter (nm)	PDI
ZnO-st NCs	(i)	65	0.28	(i)	56	0.52	ZnO-NH$_2$ st NCs	Fresh	162	0.26
	(ii)	60 *	0.38	(ii)	155	0.36		After 9 months	410	0.25
	(iii)	53 **	0.38	(iii)	108	0.18				
ZnO-mw NCs	(i)	70	0.19	(i)	52	0.12	ZnO-NH$_2$ mw NCs	Fresh	144	0.11
	(ii)	71	0.14	(ii)	56	0.20		After 9 months	105	0.11
	(iii)	64	0.18	(iii)	49	0.26				

* Peak 1: 40 nm; Peak 2: 92 nm; ** Peak 1: 28 nm; Peak 2: 58 nm

Following the in-depth morphological and structural characterization, we have focused our attention on the physico-chemical analysis, starting from the rear surface, by means of the XPS technique, as reported in Figure 6. From the survey scans (see Figure 6a for the ZnO-mw NCs, Figure 6b for ZnO-st NCs, and Figure 6c for ZnO-mw-NH$_2$ NCs) the relative atomic concentration (at.%) of each element is evaluated, as also listed in Table 2. The results on the ZnO-st-NH$_2$ NCs are not reported in view of the similarities with respect to the other functionalized sample, ZnO-mw-NH$_2$ NCs. Apart from Zn and O, we have also found C (due to the contamination from adsorbates) and N only in the functionalized ZnO-mw sample, as expected. In order to calculate the Zn/O ratio, we have subtracted from the O amount the components due to the bonds between C and O, after the deconvolution of the

C1s high-resolution (HR) peaks (not reported). Therefore, it results that the microwave-assisted NCs have a Zn/O = 0.99, while the traditional ones have a Zn/O = 0.90. Furthermore, the functionalized microwave-assisted NCs show a Zn/O = 0.83. This means that there is an increase in the O amount in the rear surface of the latest two samples, which can be easily attributed to either the synthesis or functionalization procedures. In order to verify the oxidation state of Zn and O signals, the HR curves for each sample are compared. From the Zn2p$_{3/2}$ curves (Figure 6d), no significant differences between the samples can be appreciated, since the three signals are perfectly overlapped and centered at the same binding energy (1020.9 eV), ascribed to the ZnO chemical shift [28]. Also the O1s region shows an almost perfect overlap between the three samples. From the deconvolution procedures (not reported) three components are obtained and are due to: O-Zn bond (529.7 eV), O-H bond (531.0 eV) and H$_2$O residue (532.0 eV) as already reported in literature both theoretically [29] and experimentally [30]. Moreover, we have also checked the N1s region for the functionalized ZnO-mw sample, finding out that the experimental signal (reported in the inset of Figure 6c) is due to the imine group –N= (398.6 eV; 22.3 %) and amine group –NH– (399.7 eV; 77.7%). To complete the XPS analysis fully we have also acquired the valence band (VB) signal (see Figure 6f), which can give some more information regarding the DOS region, in order to have some more hints regarding the electronic band adjustment. From XPS measurements the valence band maximum (VBM) position, related to the Fermi energy level (EF), was extracted and corresponds to the 0 eV in our binding energy scale. The linear fit (not reported) of the descending part of each spectrum towards the EF, have given these values: 2.20 eV for the ZnO-mw NCs, 2.24 eV for the traditional ZnO-st sample and 2.14 eV for the amine-functionalized ZnO-mw one. These values are in accordance with that reported in the literature by Kamarulzaman et al. [31] for nanostructured ZnO particles.

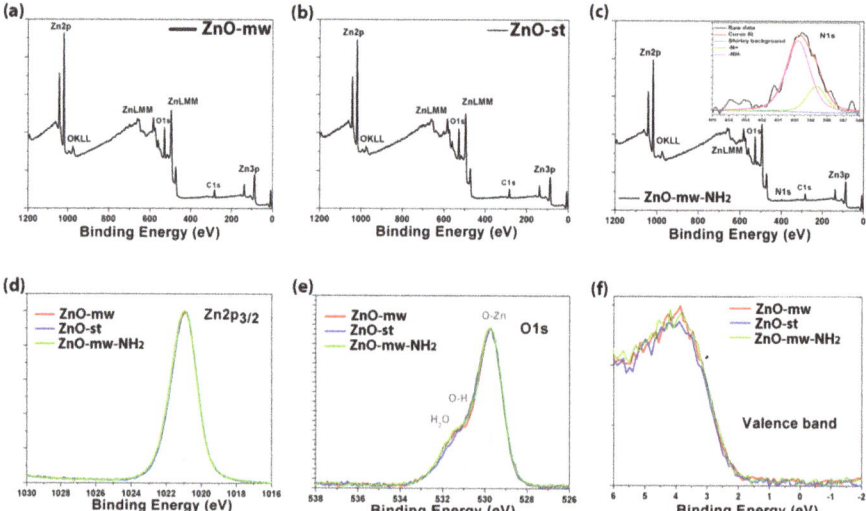

Figure 6. X-ray photoelectron spectroscopy (XPS) analysis: survey scans of (**a**) microwave-assisted synthesized ZnO nanocrystals (ZnO-mw NCs), (**b**) conventionally-synthesized NCs (ZnO-st NCs), and (**c**) amino-propyl functionalized NCs prepared by microwave assisted synthesis (ZnO-mw-NH$_2$ NCs). Inset of Figure 6c reports the high-resolution spectra of the N1s region of the ZnO-mw-NH$_2$ NCs. High-resolution (HR) curves comparing the three samples (ZnO-mw NCs in red, ZnO-st NCs in blue, ZnO-mw-NH$_2$ NCs in green) for (**d**) Zn2p$_{3/2}$ and (**e**) O1s regions. (**f**) Valence band (VB) signal related to the three samples.

To sum up, we can state that from XPS analysis the new microwave-assisted method produces NCs which are highly comparable, from the chemical and physical point of view, to those synthesized by the conventional solvothermal procedure.

Table 2. XPS relative atomic concentration (at.%) from survey data, together with Zn/O calculation in the last column. * Amount of oxygen, in atomic percent, that is associated with carbon (from survey data). ** Total amount of oxygen, from the survey, minus the amount of oxygen associated with carbon in atomic percent.

Sample	Atomic concentration (at.%)					
	C	O		N	Zn	Zn/O
ZnO-mw NCs	26.3	5.3 *	34.3 **	-	34.1	0.99
ZnO-st NCs	27.5	4.7	35.6	-	32.2	0.90
ZnO-mw-NH$_2$ NCs	25.4	1.7	38.5	2.4	32.0	0.83

3.2. Optical and Luminescent Properties of ZnO NCs

Owing to the fact that ZnO is one of the most excellent semiconductor materials, the prepared ZnO NCs are also characterized from the optical point of view. Actually, the optical and especially the luminescent properties of various ZnO nanostructures are well documented in the literature [32], both for spherical-shaped nanoparticle or nanowire form. In particular, ZnO NPs are reported to show good photophysical properties that, coupled conveniently with surface modifications, can be efficiently exploited as quantum dots in a biological environment for bio-imaging purposes [33]. In general, the sol-gel synthesis route and the large surface-to-volume ratio of the nanostructures can result in numerous defects on the surface of the ZnO NPs inducing a strong visible emission. In this regard, the optical properties of pristine ZnO NCs from both synthetic routes are investigated in this work. Furthermore, the literature reports about the influence of surface modification on the luminescence of colloidal ZnO nanoparticles [34,35].

UV−vis absorption spectroscopy was performed in the region between 300 and 800 nm, to point out the optical properties of the ZnO NCs and the related band gap (Figure 7). A comparison between the optical behaviour of NCs prepared by microwave-assisted synthesis and those synthesized by the standard one is provided, showing no differences between them. Actually, a typical and intense UV absorption is recorded for both kinds of nanocrystals in the region from 300 nm to 380 nm, characteristic of crystalline ZnO. Absence of absorption is recorded above 380 nm, showing full transparency in the visible region of the prepared NCs.

The optical band gap (Eg) of the samples was calculated using the Tauc's method from the absorption spectra, as previously reported by some of us [36], see Figure 7b. According to this method, the plot shows a linear region just above the optical absorption edge. For the investigated samples, the resulting Eg is of 3.32–3.34 eV at room temperature, thus showing almost no significant variations among them. The differences in particle size and shape between the ZnO-st and ZnO-mw nanocrystals, cannot be appreciated in these spectra and the extracted band-gap values. In particular, the heterogeneity of size distribution observed for the ZnO-st NCs is still within a few nanometers, (i.e., from 6 to 20 nm, as estimated by FESEM) and the literature even do not report differences in UV-vis spectra from even broader sizes or shape variations in ZnO nanostructures [37].

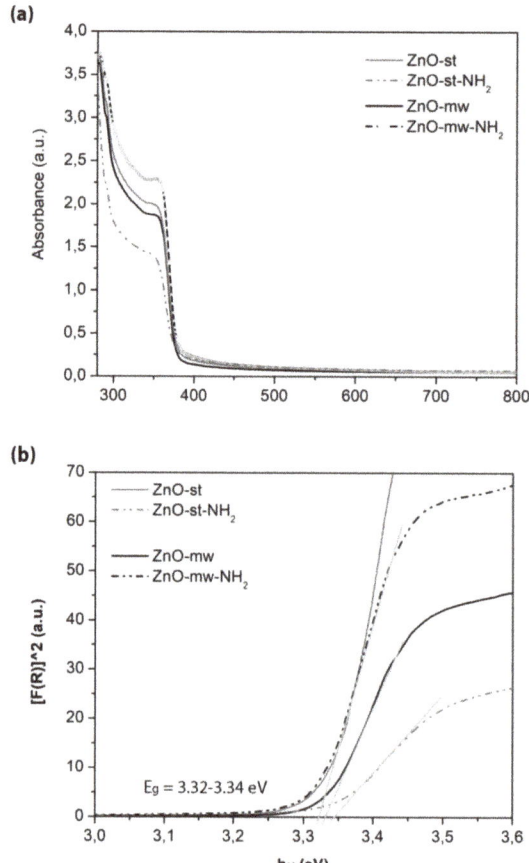

Figure 7. Characterization of the optical properties of pristine (solid lines) and functionalized (dashed lines) ZnO NCs obtained with the two preparation methods (conventional, in grey vs. microwave, in black). (**a**) Ultraviolet–visible (UV–vis) absorption spectra and (**b**) optical band gap (Eg).

The fluorescence excitation and emission spectra are reported in Figure 8, comparing the behaviour of the ZnO NCs synthesized with the two preparation methods, conventional versus microwave-assisted one, and after their functionalization with amine groups, respectively. In the fluorescence excitation spectrum (Figure 8a), the highest excitation can be observed at around 380 nm and is similar for both pristine nanocrystals (solid curves), with a slightly higher excitation peak for the ZnO-mw NCs. This excitation can be ascribed to the direct exciton transition, i.e., the excited electron recombination with holes in the valence band (VB) or in traps near the VB [32]. Furthermore, a stronger intensity is recorded after surface functionalization with amino-propyl groups (dashed curves), in particular for the ZnO-mw-NH$_2$ NCs. The enhanced fluorescence intensities of functionalized ZnO nanostructures were also previously described [34,35] and our results are in accordance with those in the literature. It is reported that amine-functionalized ZnO QDs further enhances the ZnO fluorescence by the well-known electron-donor effects of amine groups [35].

Considering the fluorescence emission spectra in Figure 8b, both kinds of NCs show a good and broad visible emission, from 500 to 700 nm approximately, when excited at λex = 380 nm. This visible emission is ascribed in the literature to crystalline defects, although these mechanisms are controversial and discussed so far. Actually, many point defects were suggested, including oxygen vacancies, oxygen interstitials, anti-site oxygen, zinc vacancies, zinc interstitials, and surface states [38]. As previously reported [39], there are two main mechanisms under discussion and considered responsible for the ZnO visible emission: (i) the recombination of an electron from the conduction band (CB) with a hole in a deep trap, and (ii) the recombination of holes from the VB with a deeply trapped electron.

It is again interesting to observe also in these emission spectra that the surface functionalization of the nanocrystals enhances their green emission, owing again to the electron-donor effects of amine groups.

Collectively, these results show that our ZnO NCs confirm previous literature data [32,39] about the good optical and luminescence properties of ZnO at the nanoscale. Furthermore, the presence of amine-functional groups leads to an enhancement of these luminescence properties, both in the excitonic emission and the green fluorescence emission. In addition, the surface functionalization with reactive amine-groups is useful for further biological modifications, i.e., anchoring of proteins and targeting ligands, as well as interaction with living cells, as reported below.

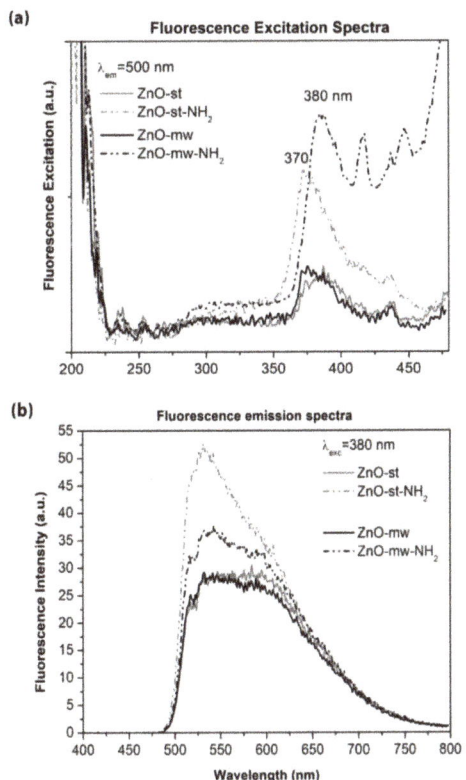

Figure 8. Luminescence properties of the pristine (solid line) and functionalized (dashed line) ZnO NCs obtained with the two preparation methods (conventional, in grey vs. microwave, in black). (**a**) Excitation spectra (recorded at the emission wavelength of 550 nm), (**b**) emission spectra (acquired at the excitation wavelength of 380 nm).

3.3. Cytotoxicity and Cell Internalization of ZnO Nanoparticles (NPs)

The cytotoxic effect and cell internalization rates of both conventional and microwave-synthesized ZnO nanocrystals was carried out on the amine-functionalized ones, i.e., ZnO-NH$_2$ NCs, against the KB cancerous human cell line. This choice was made since the amine-functionalized NCs, as previously stated, can be easily labelled with fluorescent dyes for the detection at flow cytometry or further equipped with other functional biomolecules, as previously reported [25]. Furthermore, in a prospective approach to use them as nanoimaging tools, both ZnO-NH$_2$ NCs have shown the highest luminescence properties.

The WST-1 assay was used to quantify cell viability expressed as % of control. As shown in the left panel of Figure 9, ZnO-st-NH$_2$ NCs did not exhibit any significant dose-dependent toxicity on cells. In details, the percentage of cell viability for 10, 15, 20 and 25 µg/mL concentrations result of 84% ± 12%, 89% ± 6%, 83% ± 9%, 66% ± 18%, respectively. Conversely, the results on ZnO-mw-NH$_2$ NCs show a significant decrease of KB cancer cells viability in a concentration-dependent manner, as analysed by WST-1 assay after 24 h of exposure (right panel of Figure 9). Interesting evidence is the experimental toxicity range showed by the ZnO-mw-NH$_2$ NCs; starting from 93% ± 4% at 10 µg/mL, the viability percentage decreases to 34% ± 11% at 20 µg/mL and to 23% ± 5% at 25 µg/mL. We note in details that the cell viability was significantly higher ($p \leq 0.001$) for the treatment with 10 µg/mL than with those obtained with 20 and 25 µg/mL NCs concentration treatments. Furthermore, it should be evidenced that the 15 µg/mL concentration value could represent an interesting, effective and biocompatible cut-off. In addition, after exposure to ZnO-mw-NH$_2$ NCs, KB cell line showed an IC50 value of 14 µg/mL.

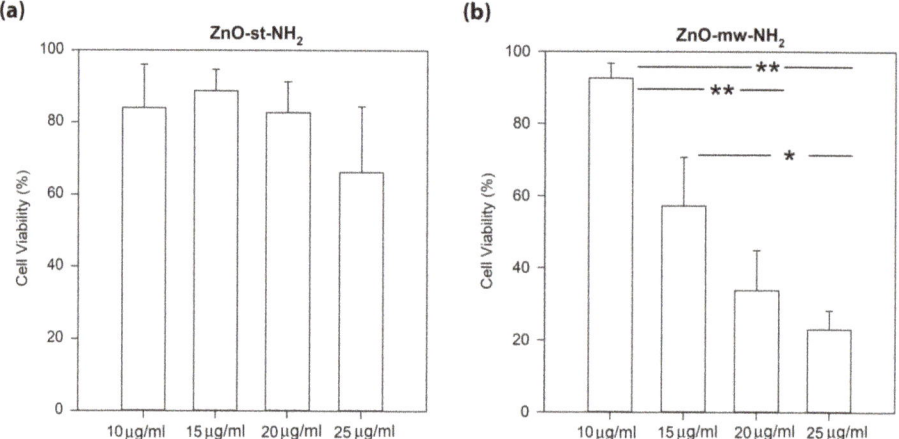

Figure 9. Comparison of the cell viability values of the (**a**) ZnO-st-NH$_2$ NCs and (**b**) ZnO-mw-NH$_2$ NCs at different concentrations toward KB cancer cell lines, detected by the WST-1 assay method. Cells were exposed in cell culture medium with different concentrations for 24 h. Results are expressed as the percent of cells viability compared to the control. The data are presented as the mean ± standard error (SE). ** $p \leq 0.001$; * $p = 0.05$.

The treatment of KB cells with either ZnO-st-NH$_2$ or ZnO-mw-NH$_2$ NCs for 24 h revealed that these NCs are non-toxic at 10 µg/mL and consequently we choose 10 µg/mL as a safe concentration for studying in vitro cellular uptake.

In flow cytometry the NCs load per cell is expressed as the number of events or intensity of the fluorescent signal associated with the labelled ZnO-NH$_2$-Atto633 NCs. The integrated fluorescent signal from single cells is measured by side scattering and interpreted as either a NCs-containing cell or NCs-free cell [40].

No statistically-significant differences appear between KB uptakes measured after 10 µg/mL of ZnO-st-NH$_2$ NCs (74 ± 9) and 10 µg/mL of ZnO-mw-NH$_2$ NCs (98 ± 0.6) treatments. It should be underlined that tests on cell treated with ZnO-st-NH$_2$ NCs are not so reproducible as the one made using ZnO-mw-NH$_2$ NCs, in fact, as shown above, standard error is considerably higher in the first case. In contrast, the uptake for cells treated with ZnO-mw-NH$_2$ NCs is reproducible and well defined as reported in the representative flow cytometry curve showed in Figure 10.

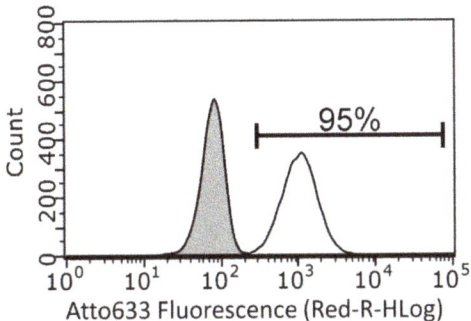

Figure 10. Representative flow cytometry curves of stained Atto 633 ZnO-mw-NH$_2$ NCs uptake in KB cells. Grey area represents the untreated cells signal while the white one was obtained by cells incubated with 10 µg/mL ZnO-mw-NH$_2$ stained with Atto633 for 24 h. Quantification of positively stained events, characterized by a shift in fluorescence intensity compared to untreated cells, was calculated setting a positive threshold beyond the grey area. Each condition is done in triplicate.

4. Conclusions

In this work we report a new ZnO microwave-assisted solvothermal synthesis, optimized for biological uses. Homogeneous and spherical nanocrystals of 20 nm were obtained, with high reaction yields, and fully characterized from the physical-chemical point of view. By comparison with a chemically equivalent wet synthetic method we were able to evaluate their different biological behaviour, in terms of cytotoxicity and cell internalization in KB cancerous human cells.

X-ray diffraction, XPS and optical analysis demonstrated similar physical-chemical properties of the ZnO NCs obtained by the two different synthetic procedures, in terms of surface chemistry and electron band gap. However, we found substantial differences related to the hydrodynamic size, shelf-life stability (tendency to agglomerate in time) leading to a better reproducibility of biological test outcomes. Through a deep statistical analysis, it was in fact possible to estimate that the ZnO NCs, obtained via microwave synthesis, show more reproducible and reliable results.

These findings suggest that not only different preparation methods, but also similar procedures that generate particles with the same surface chemistry could drive biological responses to different ways. In particular, the ability to control and to obtain narrow and reliable NPs size distributions and highly stable behaviour in solution can be considered crucial factors in drive reproducible results, i.e., cytotoxicity and cell internalization tests. In fact, minor changes within the same synthetic route can alter both the shape and the size distribution. In our case, we demonstrate how the proposed microwave procedure is highly effective to better control and optimize the ZnO NCs morphology and size. Furthermore, we prospectively welcome the use of nano-sized ZnO particles with surface modification. We actually demonstrated that amine-functionalized NCs possess improved optical properties, useful for further bio-imaging applications, and allow future biomolecules anchoring, i.e., drugs, proteins or targeting ligands against cancer cells.

Author Contributions: Conceptualization, N.G., T.L. and V.C.; Methodology, N.G., T.L., M.C., A.F. and V.C.; Validation, B.D., M.C., L.R., M.C., M.L. and A.C.; Resources, A.F., V.C.; Data Curation, N.G., T.L. and V.C.; Writing-Original Draft Preparation, N.G., T.L., A.C., M.C. and V.C.; Writing-Review & Editing, N.G., T.L., M.L., A.F. and V.C.; Supervision, N.G., T.L. and V.C.; Project Administration, V.C.; Funding Acquisition, V.C. All authors have given approval to the final version of the manuscript.

Funding: This work has received funding from the European Research Council (ERC) under the European Union's Horizon 2020 research and innovation programme (Grant Agreement No 678151–Project Acronym "TROJANANOHORSE"–ERC starting Grant).

Acknowledgments: Cristiano Di Benedetto is thankfully acknowledged for preliminary CTEM imaging.

Conflicts of Interest: There are no conflicts to declare.

References

1. Kim, D.; Shin, K.; Kwon, S.G.; Hyeon, T. Synthesis and biomedical applications of multifunctional nanoparticles. *Adv. Mater.* **2018**, *30*, 1802309. [CrossRef]
2. Zhu, P.; Weng, Z.; Li, X.; Liu, X.; Wu, S.; Yeung, K.W.K.; Wang, X.; Cui, Z.; Yang, X.; Chu, P.K. Biomedical applications of functionalized ZnO nanomaterials: From biosensors to bioimaging. *Adv. Mater. Interfaces* **2016**, *3*, 1500494. [CrossRef]
3. Racca, L.; Canta, M.; Dumontel, B.; Ancona, A.; Limongi, T.; Garino, N.; Laurenti, M.; Canavese, G.; Cauda, V. Zinc oxide nanostructures in biomedicine. In *Smart Nanoparticles for Biomedicine*; Ciofani, G., Ed.; Elsevier: Amsterdam, Netherlands, 2018; pp. 171–187.
4. Zhang, L.; Yin, L.; Wang, C.; Lun, N.; Qi, Y.; Xiang, D. Origin of visible photoluminescence of ZnO quantum dots: Defect-dependent and size-dependent. *J. Phys. Chem. C* **2010**, *114*, 9651–9658. [CrossRef]
5. Huang, X.; Zheng, X.; Xu, Z.; Yi, C. Zno-based nanocarriers for drug delivery application: From passive to smart strategies. *Int. J. Pharm.* **2017**, *534*, 190–194. [CrossRef]
6. Laurenti, M.; Cauda, V. Gentamicin-releasing mesoporous zno structures. *Materials* **2018**, *11*, 314. [CrossRef]
7. Cauda, V.; Gazia, R.; Porro, S.; Stassi, S.; Canavese, G.; Roppolo, I.; Chiolerio, A. Nanostructured ZnO materials: Synthesis, properties and applications. In *Handbook of Nanomaterial Properties*; Bhushan, B.L.D., Schricker, S.R., Sigmund, W., Zauscher, S., Eds.; Springer: Berlin/Heidelberg, Germany, 2014; Volume XXVII.
8. Garino, N.; Lamberti, A.; Gazia, R.; Chiodoni, A.; Gerbaldi, C. Cycling behaviour of sponge-like nanostructured ZnO as thin-film li-ion battery anodes. *J. Alloy Compd.* **2014**, *615*, S454–S458. [CrossRef]
9. Cauda, V.; Pugliese, D.; Garino, N.; Sacco, A.; Bianco, S.; Bella, F.; Lamberti, A.; Gerbaldi, C. Multi-functional energy conversion and storage electrodes using flower-like zinc oxide nanostructures. *Energy* **2014**, *65*, 639–646. [CrossRef]
10. Laurenti, M.; Stassi, S.; Canavese, G.; Cauda, V. Surface engineering of nanostructured ZnO surfaces. *Adv. Mater. Interfaces* **2017**, *4*, 1600758. [CrossRef]
11. Jiang, J.; Pi, J.; Cai, J. The advancing of zinc oxide nanoparticles for biomedical applications. *Bioinorg. Chem. Appl.* **2018**, 18. [CrossRef]
12. Martínez-Carmona, M.; Gun'ko, Y.; Vallet-Regí, M. ZnO nanostructures for drug delivery and theranostic applications. *Nanomaterials* **2018**, *8*, 268. [CrossRef]
13. Ali, A.; Phull, A.R.; Zia, M. Elemental zinc to zinc nanoparticles: Is ZnO NPs crucial for life? Synthesis, toxicological and environmental concerns. *Nanotechnol. Rev.* **2018**, *7*, 413–441. [CrossRef]
14. Shen, C.; James, S.A.; de Jonge, M.D.; Turney, T.W.; Wright, P.F.A.; Feltis, B.N. Relating cytotoxicity, zinc ions, and reactive oxygen in ZnO nanoparticle-exposed human immune cells. *Toxicol. Sci.* **2013**, *136*, 120–130. [CrossRef]
15. Han, L.; Zhai, Y.; Liu, Y.; Hao, L.; Guo, H. Comparison of the in vitro and in vivo toxic effects of three sizes of zinc oxide (ZnO) particles using flounder gill (fg) cells and zebrafish embryos. *J. Ocean Univ. China* **2017**, *16*, 93–106. [CrossRef]
16. Duraimurugan, J.; Kumar, G.S.; Venkatesh, M.; Maadeswaran, P.; Girija, E.K. Morphology and size controlled synthesis of zinc oxide nanostructures and their optical properties. *J. Mater. Sci. Mater. Electron.* **2018**, *29*, 9339–9346. [CrossRef]
17. Kalantari Bolaghi, Z.; Masoudpanah, S.M.; Hasheminiasari, M. Photocatalytic properties of ZnO powders synthesized by conventional and microwave-assisted solution combustion method. *J. Sol-Gel Sci. Technol.* **2018**, *86*, 711–718. [CrossRef]
18. Chaudhary, S.; Kaur, Y.; Umar, A.; Chaudhary, G.R. Ionic liquid and surfactant functionalized ZnO nanoadsorbent for recyclable proficient adsorption of toxic dyes from waste water. *J. Mol. Liq.* **2016**, *224*, 1294–1304. [CrossRef]
19. Cravotto, G.; Carnaroglio, D. *Microwave Chemistry*; Springer Book Series: Berlin, Germany; Boston, MA, USA, 2017.
20. Horikoshi, S.; Serpone, N. *Microwaves in Nanoparticle Synthesis: Fundamentals and Applications*; Wiley-VCH Verlag GmbH & Co., KGaA: Weinheim, Germany, 2013.

21. Zhu, Y.-J.; Chen, F. Microwave-assisted preparation of inorganic nanostructures in liquid phase. *Chem. Rev.* **2014**, *114*, 6462–6555. [CrossRef]
22. Quirino, M.R.; Oliveira, M.J.C.; Keyson, D.; Lucena, G.L.; Oliveira, J.B.L.; Gama, L. Synthesis of zinc oxide by microwave hydrothermal method for application to transesterification of soybean oil (biodiesel). *Mater. Chem. Phys.* **2017**, *185*, 24–30. [CrossRef]
23. Wojnarowicz, J.; Chudoba, T.; Gierlotka, S.; Lojkowski, W. Effect of microwave radiation power on the size of aggregates of ZnO NPs prepared using microwave solvothermal synthesis. *Nanomaterials* **2018**, *8*, 343. [CrossRef]
24. Wojnarowicz, J.; Chudoba, T.; Koltsov, I.; Gierlotka, S.; Dworakowska, S.; Lojkowski, W. Size control mechanism of ZnO nanoparticles obtained in microwave solvothermal synthesis. *Nanotechnology* **2018**, *29*, 065601. [CrossRef]
25. Dumontel, B.; Canta, M.; Engelke, H.; Chiodoni, A.; Racca, L.; Ancona, A.; Limongi, T.; Canavese, G.; Cauda, V. Enhanced biostability and cellular uptake of zinc oxide nanocrystals shielded with a phospholipid bilayer. *J. Mater. Chem. B* **2017**, *5*, 8799–8813. [CrossRef] [PubMed]
26. Sanginario, A.; Cauda, V.; Bonanno, A.; Sapienza, S.; Demarchi, D. An electronic platform for real-time detection of bovine serum albumin by means of amine-functionalized zinc oxide microwires. *RSC Adv.* **2016**, *6*, 891–897. [CrossRef]
27. IC50 Calculator. Available online: https://www.Aatbio.Com/tools/ic50-calculator (accessed on 23 October 2018).
28. Al-Gaashani, R.; Radiman, S.; Daud, A.R.; Tabet, N.; Al-Douri, Y. XPS and optical studies of different morphologies of ZnO nanostructures prepared by microwave methods. *Ceram. Int.* **2013**, *39*, 2283–2292. [CrossRef]
29. Kotsis, K.; Staemmler, V. Ab initio calculations of the O1s XPS spectra of ZnO and Zn oxo compounds. *Phys. Chem. Chem. Phys.* **2006**, *8*, 1490–1498. [CrossRef] [PubMed]
30. Ogata, K.; Komuro, T.; Hama, K.; Koike, K.; Sasa, S.; Inoue, M.; Yano, M. Characterization of undoped ZnO layers grown by molecular beam epitaxy towards biosensing devices. *Phys. Status Solidi* **2004**, *241*, 616–619. [CrossRef]
31. Kamarulzaman, N.; Kasim, M.F.; Rusdi, R. Band gap narrowing and widening of ZnO nanostructures and doped materials. *Nanoscale Res. Lett.* **2015**, *10*, 346. [CrossRef]
32. Zhang, Z.-Y.; Xiong, H.-M. Photoluminescent ZnO nanoparticles and their biological applications. *Materials* **2015**, *8*, 3101. [CrossRef]
33. Hong, H.; Shi, J.; Yang, Y.; Zhang, Y.; Engle, J.W.; Nickles, R.J.; Wang, X.; Cai, W. Cancer-targeted optical imaging with fluorescent zinc oxide nanowires. *Nano Lett.* **2011**, *11*, 3744–3750. [CrossRef]
34. Norberg, N.S.; Gamelin, D.R. Influence of surface modification on the luminescence of colloidal ZnO nanocrystals. *J. Phys. Chem. B* **2005**, *109*, 20810–20816. [CrossRef]
35. Li, S.; Sun, Z.; Li, R.; Dong, M.; Zhang, L.; Qi, W.; Zhang, X.; Wang, H. ZnO nanocomposites modified by hydrophobic and hydrophilic silanes with dramatically enhanced tunable fluorescence and aqueous ultrastability toward biological imaging applications. *Sci. Rep.* **2015**, *5*, 8475. [CrossRef]
36. Hernández, S.; Cauda, V.; Chiodoni, A.; Dallorto, S.; Sacco, A.; Hidalgo, D.; Celasco, E.; Pirri, C.F. Optimization of 1d ZnO@TiO$_2$ core–shell nanostructures for enhanced photoelectrochemical water splitting under solar light illumination. *ACS Appl. Mater. Interfaces* **2014**, *6*, 12153–12167. [CrossRef] [PubMed]
37. Morales-Flores, N.; Galeazzi, R.; Rosendo, E.; Díaz, T.; Velumani, S.; Pal, U. Morphology control and optical properties of ZnO nanostructures grown by ultrasonic synthesis. *Adv. Nano Res.* **2013**, *1*, 59–70. [CrossRef]
38. Özgür, Ü.; Alivov, Y.I.; Liu, C.; Teke, A.; Reshchikov, M.A.; Doğan, S.; Avrutin, V.; Cho, S.-J.; Morkoç, H. A comprehensive review of ZnO materials and devices. *J. Appl. Phys.* **2005**, *98*, 041301. [CrossRef]
39. Xiong, H.-M. Photoluminescent ZnO nanoparticles modified by polymers. *J. Mater. Chem.* **2010**, *20*, 4251–4262. [CrossRef]
40. Drasler, B.; Vanhecke, D.; Rodriguez-Lorenzo, L.; Petri-Fink, A.; Rothen-Rutishauser, B. Quantifying nanoparticle cellular uptake: Which method is best? *Nanomedicine* **2017**, *12*, 1727–1744. [CrossRef] [PubMed]

© 2019 by the authors. Licensee MDPI, Basel, Switzerland. This article is an open access article distributed under the terms and conditions of the Creative Commons Attribution (CC BY) license (http://creativecommons.org/licenses/by/4.0/).

Article

Osteogenic Effect of ZnO-Mesoporous Glasses Loaded with Osteostatin

Rebeca Pérez [1], Sandra Sanchez-Salcedo [1,2], Daniel Lozano [1,2], Clara Heras [1], Pedro Esbrit [1,3], María Vallet-Regí [1,2] and Antonio J. Salinas [1,2,*]

[1] Departamento de Química en Ciencias Farmacéuticas, Facultad de Farmacia, Universidad Complutense de Madrid, UCM, Instituto de Investigación Hospital 12 de Octubre, imas12, 28040 Madrid, Spain; rebecapr93@gmail.com (R.P.); sansanch@ucm.es (S.S.-S.); danlozan@ucm.es (D.L.); claheras@ucm.es (C.H.); pesbrit@gmail.com (P.E.); vallet@ucm.es (M.V.-R.)
[2] Networking Research Center on Bioengineering, Biomaterials and Nanomedicine (CIBER-BBN), 28040 Madrid, Spain
[3] Instituto de Investigación Sanitaria (IIS)-Fundación Jiménez Díaz, 28040 Madrid, Spain
* Correspondence: salinas@ucm.es; Tel.: +34-913-941-790; Fax: +34-913-941-786

Received: 5 July 2018; Accepted: 1 August 2018; Published: 4 August 2018

Abstract: Mesoporous Bioactive Glasses (MBGs) are a family of bioceramics widely investigated for their putative clinical use as scaffolds for bone regeneration. Their outstanding textural properties allow for high bioactivity when compared with other bioactive materials. Moreover, their great pore volumes allow these glasses to be loaded with a wide range of biomolecules to stimulate new bone formation. In this study, an MBG with a composition, in mol%, of 80% SiO_2–15% CaO–5% P_2O_5 (Blank, BL) was compared with two analogous glasses containing 4% and 5% of ZnO (4ZN and 5ZN) before and after impregnation with osteostatin, a C-terminal peptide from a parathyroid hormone-related protein ($PTHrP_{107-111}$). Zn^{2+} ions were included in the glass for their bone growth stimulator properties, whereas osteostatin was added for its osteogenic properties. Glasses were characterized, and their cytocompatibility investigated, in pre-osteoblastic MC3T3-E1 cell cultures. The simultaneous additions of osteostatin and Zn^{2+} ions provoked enhanced MC3T3-E1 cell viability and a higher differentiation capacity, compared with either raw BL or MBGs supplemented only with osteostatin or Zn^{2+}. These in vitro results show that osteostatin enhances the osteogenic effect of Zn^{2+}-enriched glasses, suggesting the potential of this combined approach in bone tissue engineering applications.

Keywords: mesoporous glasses; ZnO-additions; osteostatin loading; osteosteoblast cell cultures; osteogenic effect

1. Introduction

Bone regeneration is a natural event, but there are certain clinical situations where this physiological process is impaired. For instance, when either the bone defect to be repaired is too large or bone has lost its regenerative capacity as occurs in osteoporosis conditions. In these cases, bone regeneration needs to be stimulated by using bone tissue engineering approaches [1,2]. Such approaches use constructs formed by 3-D porous scaffolds decorated with biological signals and/or bone-forming cells. In the last decade, SiO_2–CaO–P_2O_5 mesoporous bioactive glasses (MBGs) were proposed as optimum candidates for these scaffolds. These glasses exhibit bone regenerative properties and highly ordered mesoporous structures enabling binding and release bone promoting agents [1,3]. Moreover, the huge surface area and pore volumes of MBGs yield quicker in vitro responses when compared with other bioactive materials [4,5]. The behaviour of these glasses in a biological medium can be improved by incorporating bioactive metal ions in the glass network. This is the case of Zn^{2+} ions which exhibit osteogenic and angiogenic features, as well as antioxidant, cancer preventive, and antimicrobial activities [6–10]. In this regard, since bacterial infection [11] is an important problem after bone implant surgery [12,13], the combination of the regenerative properties of MBGs with the beneficial effects of Zn^{2+} ions has generated potential interest in bioengineering applications [14].

Following the bone tissue engineering principles, the bioactivity of a scaffold can be improved by loading it with osteogenic agents, such as parathyroid hormone (PTH)-related protein (PTHrP), which is emerging as an interesting promoter of bone regeneration. PTHrP contains an N-terminal 1–37 region homologous to PTH and a C-terminal PTH-unlike region containing the highly conserved 107–111 sequence osteostatin [15]. N-terminal PTHrP analogues have been shown to induce bone anabolism in rodents and humans upon systemic intermittent administration [16,17]. On the other hand, osteostatin has anti-resorptive activity [18], but also exhibits osteogenic features in vitro and in vivo [19–24]. Moreover, it has recently been shown that osteostatin coating onto various types of ceramic implants accelerates healing of critical and noncritical bone defects in the long bones of adult normal and osteoporotic rabbits and in rats [25–29]. Therefore, recent findings point to osteostatin as an attractive small peptide for consideration in a bone tissue engineering scenario.

In this study, the biological consequences of the concurrent inclusion of ZnO and osteostatin impregnation in MBGs were investigated. Three MBGs were synthesized, all with a basic composition of 80% SiO_2–15% CaO–5% P_2O_5 (mol%), containing or not (Blank, BL) 4% or 5% ZnO, respectively (4ZN and 5ZN). These compositions were selected based on our previous studies [14], which were consistent with those reported for other glass systems showing 5% as the maximum content of ZnO enhancing osteoblast cell development without being cytotoxic [30–32]. MBG powders were processed as disk-shape pieces for several in vitro studies: uptake and release of osteostatin; assays in simulated body fluid (SBF); release of the inorganic ions, calcium, phosphate, and zinc from disks to the surrounding medium; and bioactivity in mouse pre-osteoblastic MC3T3-E1 cell cultures. This approach allowed us to evaluate the putative advantage of loading osteostatin onto ZnO-containing glasses to produce an optimal biomaterial for bone regeneration.

2. Experimental

2.1. Synthesis of the MBGs as Powders and Processing into Disks

The synthesis of the MBGs was made through the EISA (Evaporation-Induced Self Assembly) method, using 4.5 g Pluronic® P123 as surfactant, 85 mL ethanol (99.98%), as solvent, and 1.12 mL 0.5 N HNO_3 as catalyst. The process was carried out for 1 h under stirring at 250 rpm, covering the flask with Parafilm® to prevent the solvent evaporation. Then, the appropriate amounts of tetraethyl orthosilicate (TEOS), $Ca(NO_3)_2 \cdot 4H_2O$, triethyl phosphate (TEP) and $Zn(NO_3)_2 \cdot 6H_2O$ were added as SiO_2, CaO, P_2O_5 and ZnO sources, respectively (all reagents from Sigma-Aldrich, St. Louis, MO, USA). Thus, 8.9 mL TEOS were slowly added for 3 h, followed by the addition of 0.71 mL TEP for another 3 h period. Next, 1.10 g $Ca(NO_3)_2 \cdot 4H_2O$ and the required amounts of $Zn(NO_3)_2 \cdot 6H_2O$ depending on the designed ZnO content (0.60 g for 4ZN or 0.75 g for 5ZN) were also added. The solution was continuously stirred at 250 rpm during the synthesis process. The solution was left overnight (14 h), then it was distributed in Petri dishes (30 mL/plate), and let the ethanol to evaporate at 25 °C for 7 day. Thereafter, the resulting transparent membrane was withdrawn and heated for 6 h at 700 °C (with a heating ramp of 1 °C/min). Finally, materials were gently milled on a glass mortar to prevent deterioration of the mesoporous order and sieved through a 40 µm mesh. For the in vitro assays, the powders were conformed into disks (6 mm diameter, 2 mm height) obtained by compacting 70 mg of MBG powders with 5 MPa of uniaxial pressure.

2.2. Physicochemical Characterization of Samples

The samples were characterised by CHN elemental analysis in a Macroanalyser Leco CNS-2000-I (Saint Joseph, MI, USA); Thermogravimetric and Differential Thermal analysis (TG/DTA) in the 30 °C to 900 °C interval (air flow: 100 mL/min) in a Perkin Elmer iPyris Diamond system r

(Waltham, MA, USA), Fourier transformed infrared (FTIR) spectroscopy in a Thermo Scientific Nicolet iS10 apparatus (Waltham, MA, USA) equipped with a SMART Golden Gate attenuated total reflection ATR diffuse reflectance accessory; Small-Angle X-ray diffraction, SA-XRD, in a X'pert-MPD system (Eindhoven, The Netherlands) equipped with Cu Kα radiation in the 0.6 to 8° 2θ range and Transmission Electron Microscopy (TEM), in a JEM-2100 JEOL microscope operating at 200 kV (Tokyo, Japan). Samples were ultrasonically dispersed in n-butanol and deposited in a copper grid coated with a holed polyvinyl-formaldehyde layer for TEM analysis.

Moreover, samples were characterised by nitrogen adsorption and solid-state nuclear magnetic resonance (NMR). Nitrogen porosimetry was performed in a Micromeritics ASAP 2020 (Norcross, GA, USA). Samples were previously degassed 24 h at 120 °C under vacuum. The surface areas were calculated by the Brunauer-Emmett-Teller (BET) method [33], and the pore size distributions by the Barret–Joyner–Halenda (BJH) method [34]. Surface functionalization was studied by solid state single pulse magic angle spinning nuclear magnetic resonance (SP MAS NMR). The ^{29}Si and ^{31}P spectra were obtained on a Bruker Avance AV-400WB spectrometer (Karlsruhe, Germany) equipped with a solid state probe using a 4 mm zirconia rotor and spun at 10 kHz for ^{29}Si and 6 kHz in the case of ^{31}P. Spectrometer frequencies were set at 79.49 and 161.97 MHz for ^{29}Si and ^{31}P, respectively. Chemical shift values were referenced to tetramethylsilane (TMS) for ^{29}Si and H_3PO_4 ^{31}P. The time period between accumulations were 5 and 4 s for ^{29}Si and ^{31}P, respectively, and the number of scans was 10,000.

2.3. In Vitro Studies

In vitro tests were carried out in MBGs disks sterilized for 20 min under UV radiation (10 min/face) in a laminar flux cabinet. The disks maintained their stability without crumbling even for soaking times as long as 21 days in the assays performed in SBF.

2.3.1. Adsorption and Release of Osteostatin

For the adsorption assay, the disks in 24-well plates were incubated with 1 mL of phosphate-buffered saline (PBS), pH 7.4, containing or not (control) 100 nM osteostatin. Samples were left under stirring at 400 rpm, at 4 °C. Osteostatin adsorption in each type of tested MBG after 24 h was calculated based on the peptide removed from the liquid medium; whereas osteostatin release was measured by soaking the peptide-loaded disks for different times (1, 2, 24, 48, 72, and 96 h) in PBS also under stirring, at 37 °C. The amount of osteostatin in PBS medium was measured by UV spectrometry at 280 nm using a NanoDrop ND-2000 (NanoDrop Technologies, Thermo Fisher Scientific, Wilmington, DE, USA).

2.3.2. Assays in SBF

In vitro bioactivity tests were carried out by soaking the disks for 6 h, 24 h, 3 days, 7 days, 14 days, and 21 days in SBF, pH 7.4, at 37 °C [35]. SBF was previously filtered through a 0.22 μm filter to prevent bacterial contamination. The disks were placed in polyethylene flasks containing 13 mL of SBF, according to the equation Vs = Sa/10 (being Vs the SBF volume in mL and Sa the external surface area of the disks in mm^2). Inside the SBF, the disks were located in a vertical position by including them in "baskets" fabricated with platinum wire. Two replicas by material and time and a control with only SBF were included.

Before and after the assays, disks were characterized by wide angle X-ray diffraction (XRD, 2θ from 10–70) in an X'Pert-MPD (Philips) system, FTIR spectroscopy in a Thermo Scientific Nicolet iS10 (KBr pellet method), and SEM in a JSM-6400 (JEOL) microscope (Tokyo, Japan) coupled with an EDX spectroscopy system (Oxford Instruments, Abingdom, UK). Moreover, changes in Ca^{2+} concentration and pH of the liquid medium were assessed with an ILyte® electrode ion selective system (Diamond Diagnostics, Holliston, MA, USA). An in vitro bioactive behaviour in SBF is generally identified by the deposition on the material surface of amorphous calcium phosphate (ACP) layer that later on crystallized as hydroxycarbonate apatite (HCA) nanocrystals analogous to those in bone [36,37].

2.3.3. Ions Release from Disks

The release of ions was investigated by soaking the MBG disks in 2 mL of Dulbecco's modified Eagle medium (DMEM) (Sigma-Aldrich. St. Louis, MO, USA) supplemented with 10% fetal bovine serum (FBS) and antibiotics (100 U mL^{-1} penicillin, 100 mg mL^{-1} streptomycin) (usually called "complete medium") at 37 °C for different times between 24 h and 5 days. For each disk sample, the cumulative amounts of Ca, P and Zn released to the complete medium was determined by inductively coupled plasma/optical spectrometry (ICP/OES) using an OPTIMA 3300 DV device (Perkin Elmer). The concentration of each ion was determined from three replicates on the same solution split into two independent experiments.

2.3.4. Culture Cell Studies

Cell culture experiments were performed using the mouse pre-osteoblastic MC3T3-E1 cell line (subclone 4, CRL-2593; ATCC, Mannassas, VI) [25,27]. The different disks tested were placed into 6 and 24-well plates before cell seeding at 20,000 cells/cm^2 in 2 mL of α-minimum essential medium containing 10% FBS, 50 µg/mL ascorbic acid, 10 mM β-glycerol-2-phosphate, and 1% penicillin–streptomycin at 37 °C in a humidified atmosphere of 5% CO_2, and incubated for different times between 1 and 13 days. As controls, wells without disks were used. The medium was replaced every other day.

Cell numbers were determined using the CellTiter 96® AQueous Assay (Promega, Madison, WI, USA), a colorimetric method for determining the number of living cells in the culture. Cells were cultured without (control) or with the tested disks for 10 h (only measured on the disks surface) or for 2 and 5 days (measured in both the disks and well surface). Next, 40 µL of CellTiter 96® AQueous One Solution Reagent [containing 3-(4,5-dimethythizol-2-yl)-5-(3-carboxymethoxyphenyl)-2-(4-sulfophenyl)-2H-tetrazolium salt (MTS) and an electron coupling reagent (phenazine ethosulfate) that allows its combination with MTS to form a stable solution for 4 h] was added to each well (200 µL) in contact with the cells. The quantity of formazan product as measured by the amount of 490 nm absorbance (in a Unicam UV-500 spectrophotometer) is directly proportional to the number of living cells in culture. In addition, in some cell cultures, at 2 days, cells were trypsinized and counted in a hemocytometer to determine cell death by Trypan blue exclusion.

Alkaline phosphatase (ALP) activity was measured in MC3T3-E1 cell extracts obtained with 0.1% Triton X-100 at day 5 of culture, using p-nitrophenylphosphate as substrate, as previously described [38]. ALP activity was normalized to the cell protein content, determined by the Bradford's method using bovine serum albumin as standard.

Matrix mineralization was measured in MC3T3-E1 cell cultures by alizarin red staining, as described [26]. After incubation with the different disks for 12 days, cells were washed with PBS, and fixed with 75% ethanol for 1 h at room temperature. Cell cultures were stained with 40 mM alizarin red (pH 4.2) for 10–30 min at room temperature. Then, cells were washed with distilled water, and the stain was dissolved with 10% cetylpyridinium chloride in 10 mM PBS and measuring absorbance at 620 nm in a Unicam UV-500 spectrophotometer (ThermoSpectronic, Cambridge, UK).

Total RNA was isolated from MC3T3-E1 cells by a standard procedure (Trizol, Invitrogen, Groningen, The Netherlands), and gene expression was analysed by real-time PCR using a QuantStudio 5 Real-Time PCR System (Thermo Fisher Scientific, Wilmington, DE, USA). Real-time PCR was done using mouse-specific primers and TaqMan^{MGB} probe for Runx2 (Assay-by-Design^{SM}, Applied Biosystems, CA, USA). The mRNA copy numbers were calculated for each sample by using the cycle threshold (C_t) value. Glyceraldehyde 3-phosphate dehydrogenase (GAPDH) rRNA (a housekeeping gene) was amplified in parallel with Runx2. The relative gene expression was represented by $2^{-\Delta\Delta Ct}$, where $\Delta\Delta Ct = \Delta Ct_{target\ gene} - \Delta Ct_{GAPDH}$. The fold change for the treatment was defined as the relative expression compared with control, calculated as $2^{-\Delta\Delta Ct}$, where $\Delta\Delta Ct = \Delta C_{treatment} - \Delta C_{control}$ [21]. Runx2 (Mm00501578_m1; NM_001146038.2). GAPDH (Mm99999915_g1; NM_001289726.1).

Cell morphology was studied in disks using an Eclipse TS100 inverted optical microscope (Nikon) (Amsterdam, The Netherlands) after 24 h. Fluorescence microscopy was also carried out for the observation of attached cells onto the disks. After samples were fixed and permeabilized, they were incubated with Atto 565-conjugated phalloidin (dilution 1:40, Molecular Probes, Sigma-Aldrich, St. Louis, MO, USA), which stains actin filaments. Then, samples were washed with PBS and the cell nuclei were stained with l M diamino-20-phenylindole in PBS (DAPI) (Molecular Probes). Fluorescence microscopy was performed with an EVOS FL Cell Imaging System (Waltham, MA, USA) equipped with tree Led Lights Cubes (kEX (nm); kEM (nm): DAPI (357/44; 447/60), RFP (531/40; 593/40) from AMG (Advance Microscopy Group, Bothell, WA, USA).

2.3.5. Statistical Analysis

Results are expressed as mean ± SEM (SEM: standard error of mean). Statistical evaluation was carried out with nonparametric Kruskal-Wallis test and post-hoc Dunn´s test, when appropriate. A value of $p < 0.05$ was considered significant.

3. Results and Discussion

3.1. Glass Powders Characterization

Prior to obtaining t, he disks that were used for the in vitro tests, the MBG powders were characterized by several experimental techniques. CHN elemental analysis, TG/DTA, and FTIR spectroscopy showed the successful synthesis of glasses confirming the entire removal of surfactant and nitrate groups coming from Ca^{2+} and Zn^{2+} sources, and the MBG stabilization under ambient conditions after the last step of synthesis, namely the treatment at 700 °C. In addition, MBGs powders were characterised by SA-XRD and TEM, to assess if they exhibited ordered mesoporosity, and by nitrogen adsorption to determine their textural properties, i.e., specific surface area and porosity.

Figure 1A shows the SA-XRD patterns of BL, 4ZN, and 5ZN powders. As is observed, the BL pattern exhibits a sharp diffraction maximum at 1.3, indicative of mesoporous order, and a shoulder at around 2.0 in 2θ. According to our previous studies, the sharp maximum was assigned to the (10) reflection of a 2-D hexagonal phase formed by the mesopores arrangement and the shoulder to the low intensity (11) and (20) reflections of this phase [39]. In contrast, in the 4ZN and 5ZN patterns, only a diffuse maximum and shoulders at about 1.3 in 2θ were observed. This type of pattern is generally present in samples exhibiting worm-like order [40].

Figure 1B shows the high resolution TEM images of the MBG powders. BL and 4ZN images mainly show ordered areas confirming the presence of a mesoporous ordered structure. In addition, in these samples minority regions with disordered worm-like structures are present. In the TEM image of 5ZN, most of the observed areas exhibited worm-like order. Thus, TEM results confirmed those obtained by SA-XRD, demonstrating that the order of mesopores decreased with the presence of Zn^{2+} ions in the glass network.

To assess whether this decrease in the mesoporous order by Zn^{2+} ions was accompanied by a significant variation in the textural parameters, the MBG powders were characterized by nitrogen adsorption. As observed in Figure 1C, the isotherms of the three samples were type IV, characteristic of mesoporous materials. Moreover, the curves exhibit a type H1 cycle of hysteresis, indicative of the presence of cylindrical pores opened at both ends. Thus, BL, 4ZN, and 5ZN exhibited analogous features in terms of the type and shape of the pores present. The aforementioned textural properties of these glasses were then calculated from the isotherms. As seen in Figure 1C, inset, only moderate decreases took place in the textural properties as consequence of the inclusion of Zn^{2+} ions in the glass. Thus, the specific surface area of 372 m^2/g of BL slightly decreased to 362 and 340 m^2/g in 4ZN and 5ZN, respectively. Furthermore, the pore volume also experienced a moderate decrease from 0.47 to 0.38 and 0.39 cm^3/g, and the average pore diameter from 4.7 to 4.2 and 4.3 nm, respectively, in Zn-containing glasses.

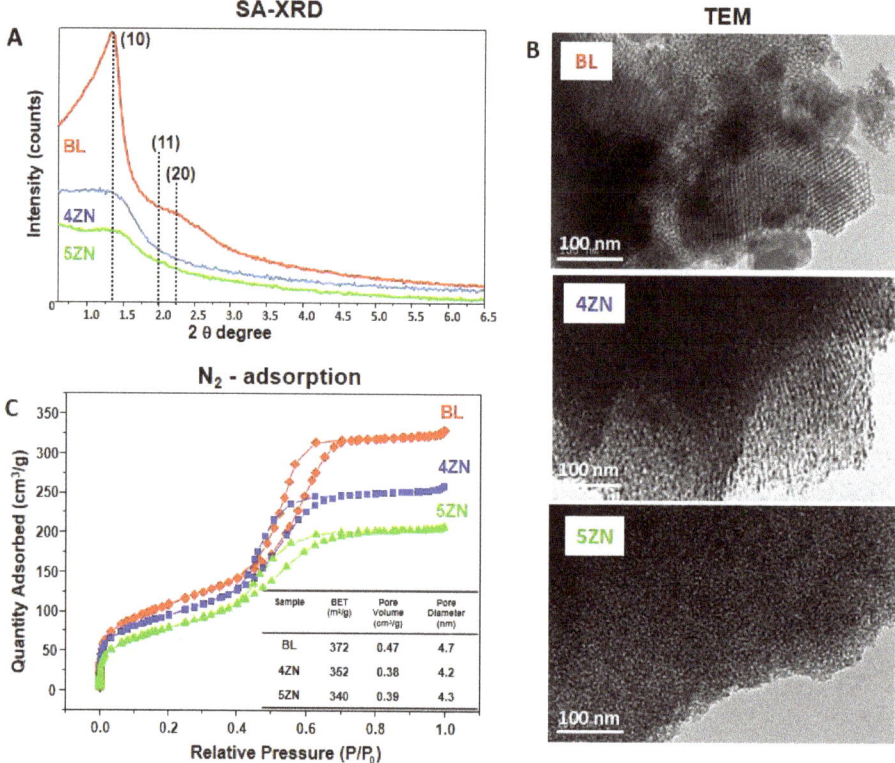

Figure 1. Physicochemical characterization of Zn-free (BL) and Zn-substituted (4ZN and 5ZN) MBG powders by: (**A**) SA-XRD; (**B**) TEM; and (**C**) N$_2$ adsorption. Inset table: calculated textural properties, i.e., specific surface area (S$_{BET}$), volume of pores (V$_P$), and pore diameter (D$_P$).

^{29}Si and ^{31}P solid state MAS NMR measurements were carried out to investigate the environments of the network formers and network modifiers species at atomic level in the MBGs (Figure 2). The NMR analysis will be related later with the release of Zn^{2+} ions in the in vitro assays with cells. In the Figure, Q^2, Q^3, and Q^4 represent, respectively, the silicon atoms (denoted Si*) in (NBO)$_2$Si*–(OSi)$_2$, (NBO)Si*–(OSi)$_3$, and Si*(OSi)$_4$ (NBO = nonbonding oxygen) [41], whereas Q^0 and Q^1, represent respectively the phosphorus atoms (denoted P*) in the PO$_4$$^{3-}$ species, (NBO)$_3$P*–(OP) and (NBO)$_2$-P*–(OP)$_2$ (NBO relative to another P atom). The chemical shifts, de-convoluted peak areas, and silica network connectivity <Q^n> for each glass composition are collected in Table 1.

Table 1. Chemical shifts (CS) and relative peak areas of MBGs obtained by ^{29}Si and ^{31}P NMR. Areas of the Q^n units were calculated by Gaussian deconvolution, the relative populations were expressed as % and the full width at half maximum, FWHM, was also included.

Sample	^{29}Si									<Q^n>	^{31}P					
	Q^4			Q^3			Q^2				Q^0			Q^1		
	CS ppm	Area (%)	FWHM ppm	CS ppm	Area (%)	FWHM ppm	CS ppm	Area (%)	FWHM ppm		CS ppm	Area (%)	FWHM ppm	CS ppm	Area (%)	FWHM ppm
BL	−112	61.7	8.04	−103	33.8	9.4	−92	4.4	18.5	3.57	1.5	91.2	5.7	−5.3	8.8	13.0
4ZN	−112	68	10.6	−102	18.9	6.0	−94	12.9	10.3	3.55	2.0	93.7	7.7	−6.9	6.3	6.1
5ZN	−110	75.7	11.8	−103	15.4	4.3	−93.5	8.92	11.6	3.67	2.4	92.9	5.8	−7.5	7.0	7.4

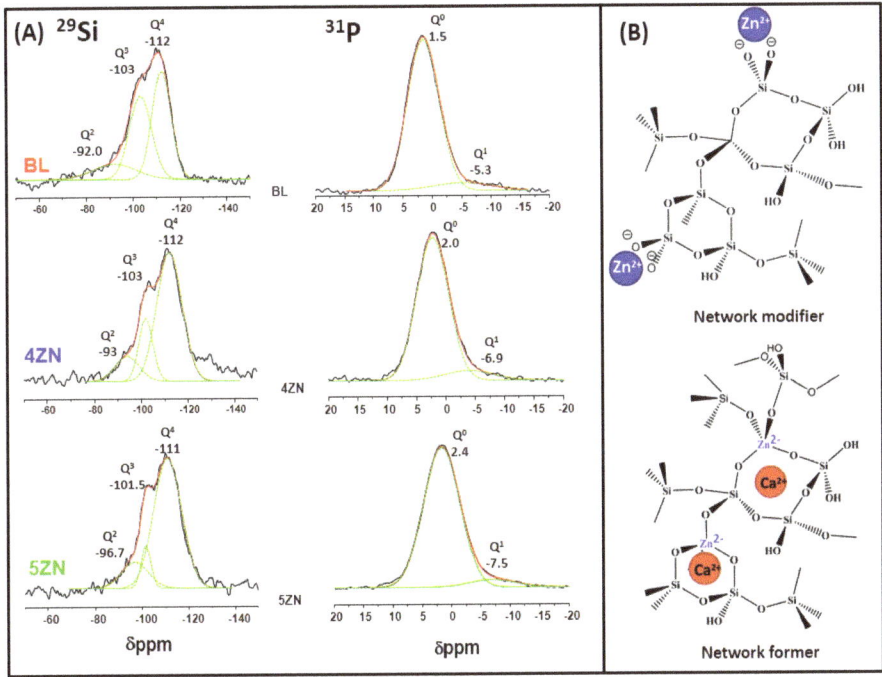

Figure 2. (A) Solid-state ^{29}Si single-pulse and ^{31}P single-pulse MAS-NMR spectra of BL, 4ZN and 5ZN. Q^n unit areas were calculated by Gaussian line-shape deconvolution and displayed in green; (B) Schematic view of Q^2 Zn and Q^4 Zn assignments from ^{29}Si MAS NMR.

In the ^{29}Si NMR spectra, the signals at −110 to −112 ppm region were assigned to Q^4; at −101 to −103 ppm to Q^3; and at −92 ppm to −96 ppm to Q^2. The BL sample was characterized by a high percentage of Q^4 and Q^3 species, and the network connectivity, $<Q^n>$, calculated for this sample was 3.57. This value is lower than reported for a sol-gel glass with identical composition that was 3.75 [40]. The relatively low values of $<Q^n>$ in MBGs is one of their features that can explain the quick in vitro bioactive response of this family of glasses.

As is observed in the Table, the inclusion of 5.0% ZnO produced an increase of $<Q^n>$, 3.67, compared with BL whereas in 4ZN a slight decrease, 3.55, was observed. These results were explained considering that when a 4.0% of ZnO was added, Zn^{2+} ions behaved as network formers with tetrahedral coordination [ZnO$_4$] which exhibit negative charge (2−). These tetrahedra attract Ca^{2+} ions that accordingly behave as charge compensators instead of as network modifiers (Figure 2B). Regarding 5ZN, the higher percentage of Q^4 species in this sample indicates a decrease in the NBO which supposes a higher contribution of Zn^{2+} as the network former compared with 4ZN, explaining the highest value of $<Q^n>$ for 5ZN. However, the amounts of ZnO were not high enough to increase substantially the depolymerisation of network in 4ZN and 5ZN with respect to BL, explaining the similar values of $<Q^n>$ obtained for the three samples (Table 1) that agree with previously reported for BL and 4ZN [14]. The increasing FWHM was due to a larger distribution of isotropic values of the chemical shift, with is caused by a decreasing short-range order of the framework structure [42]. The tetrahedral symmetry of the Q^4 units in BL sample respect to 4ZN and 5ZN samples indicates an isotropic structure when no zinc was added to the MBGs. In addition, the crystallinity of Q^3 and Q^2 were slightly greater when zinc was present in the samples.

On the other hand, the ^{31}P NMR spectra show a maximum of ≈2 ppm assigned at the Q^0 environment of amorphous orthophosphate (Figure 2A) and a second weak signal from −5.2 ppm to −7.5 ppm when the ZnO % in MBGs increases [43]. This resonance fell in the range of Q^1 tetrahedra and can be assigned to P–O–Si environments as previously reported [44,45]. Thus, P was mainly present as orthophosphate units but Zn inclusion caused a slight decrease of Q^1 units percentage, and its chemical shift pass from −5.2 ppm for BL to −7.5 ppm for 5ZN, suggesting a partial conversion of P–O–Si units into P–O–Zn units due to Zn^{2+} ion acting as a network former with more anisotropic structure than Q^1 of the BL sample. The formation of P–O–Zn was proposed for bioactive melt glasses where a shift towards lower ppm was detected when the ZnO % in the glass increased [46].

In summary, the characterizations of the MBGs powders has shown the decrease of the order of mesopores when ZnO in glasses increased. However, the textural properties of ZnO-containing MBGs remained similar to un-doped MBG (BL) with values of surface area and porosity higher than conventional sol-gel glasses [47]. Moreover, the NMR results allowed for understanding the release of Zn^{2+} ions during the in vitro assays with cells.

3.2. Textural Properties of MBG Disks

As previously mentioned, the disks used for the in vitro tests were obtained by compacting the powders at 5 MPa, then it was necessary to characterize the MBG disks by nitrogen adsorption to evaluate the textural properties after the processing. As observed in the top-left of Figure 3, the isotherms showed identical features to the MBG powders shown in Figure 1C. Moreover, at the top-right of Figure 3, the corresponding pore size distributions are shown.

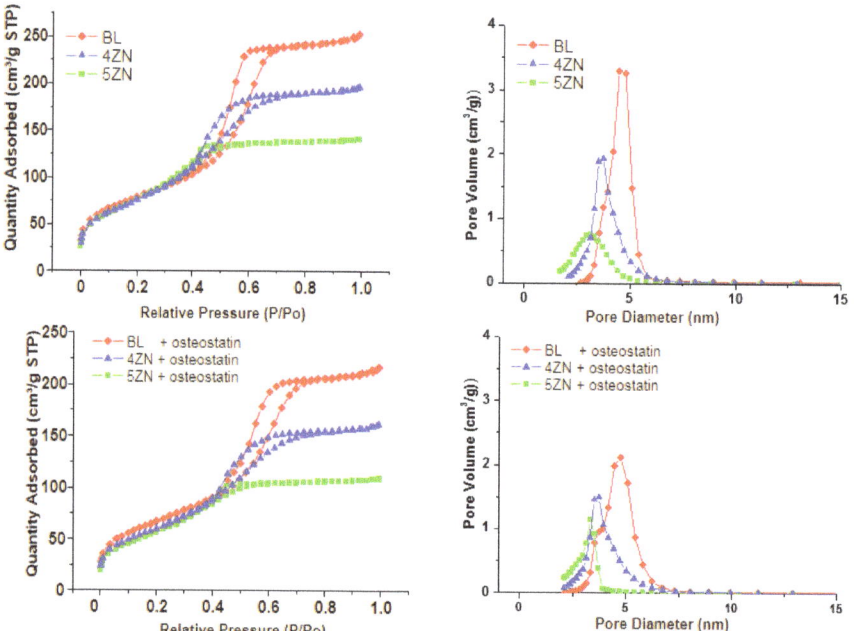

Figure 3. N_2 adsorption-desorption isotherms (left) and pore size distribution (right) of BL, 4ZN, and 5ZN disks before (upper panels) and after (lower panels) being loaded with osteostatin.

Table 2 allows for the comparison of the textural properties of MBG powders and disks. As it is observed, the disks exhibited moderate decreases of the textural properties compared with the corresponding powders. Thus, S_{BET} values, between 372 and 340 m^2/g in powders, decreased to values in the range 287–280 m^2/g in the disks, and the pore volume decreased from 0.47–0.38 cm^3/g to 0.38–0.22 cm^3/g. Therefore, textural propertied of disks remained high enough to host osteostatin molecules. The textural properties of disks after being loaded with osteostatin are also included in Table 2. As is observed, additional decreases of the specific surface area and pore volume were detected, confirming the loading of the osteostatin into the MBGs. Finally, the composition of disks was determined by EDX obtaining the values shown at the right of the Table 2. These values showed a good agreement with the nominal composition of the glasses included between brackets in the table.

Table 2. Textural properties of the glasses as powders, disks, and disks loaded with osteostatin. (S_{BET}: specific surface area; V_T: pore volume; D_P: pore diameter). On the right, experimental compositions of samples determined by EDX and nominal compositions indicated between brackets are given.

	Powders			Disks			Disks + Osteostatin			Composition (EDX) Atomic %			
	S_{BET} (m^2/g)	V_T (m^3/g)	D_P (nm)	S_{BET} (m^2/g)	V_T (m^3/g)	D_P (nm)	S_{BET} (m^2/g)	V_T (m^3/g)	D_P (nm)	SiO$_2$	CaO	P$_2$O$_5$	ZnO
BL	372	0.47	4.7	287	0.38	4.7	244	0.32	4.6	77.0 (80)	6.1 (5)	16.8 (15)	–
4ZN	352	0.38	4.2	280	0.30	3.7	221	0.24	3.6	74.3 (77)	7.0 (4.8)	14.3 (14.4)	4.2 (4)
5ZN	340	0.39	4.3	285	0.22	3.3	208	0.17	3.1	68.1 (76)	9.2 (4.8)	17.3 (14.3)	5.3 (5)

3.3. Uptake and Release of Osteostatin

After soaking the MBG disks in a 100 nM solution of osteostatin in PBS for 24 h, the mean uptake of the peptide was 63% (BL), 70% (4ZN), and 71% (5ZN) (Figure 4A), equivalent to 0.8, 0.9 and 0.95 µg/g per disk, respectively.

On the other hand, the osteostatin released from the loaded disks to the medium after 1 h was 73% (BL), 67% (4ZN), and 68% (5ZN). After 24 h, it was 95% (BL) or 90% (4ZN and 5 ZN) and it was virtually 100% for the three MGBs at 96 h (Figure 4B). It is pertinent to mention here that minimum amounts of this peptide (even in the sub-nM range) were efficient to induce osteogenic activity [26–28].

As previously told, the effect of loading osteostatin in the textural properties of disks was determined. Figure 3 includes the N$_2$ adsorption isotherms (bottom-left) and pore size distributions (bottom-right) of osteostatin-loaded disks. As is observed in Table 2, a slight decrease in the surface area and porosity was observed in the MBG disks as a consequence of osteostatin loading. These results suggest that a part of the peptide loading took place inside the pores, but without affecting the ordering of the mesopore channels.

If we assume that the release mechanism of osteostatin was diffusion through the mesopores and considering the low solubility of the glasses at the medium pH (7.4), the peptide release could be described by a deviation from the theoretical first-order behaviour of the Noyes–Whitney equations as described by Equation (1) [48,49]:

$$W_t/W_0 = A(1 - \exp^{k_1 \cdot t}) \qquad (1)$$

where W_t stands for the peptide mass released at time t; W_0 represents the maximum initial mass of the peptide inside the pores; A is the maximum amount of peptide released; and k_1 is the release rate constant, which is independent of peptide concentration and gives information about the solvent accessibility and the diffusion coefficient through mesoporous channels.

This model was successfully applied for the release of different drugs from insoluble mesoporous matrices with a similar structure [48]. According to this model, peptide release is faster within the first 24 h, reaching a stationary phase after 48 h. This deviation could be due to several factors, such as the peptide volume, the distortion of the mesopore channels, and/or the release of peptide molecules adsorbed on the external surface of the matrices.

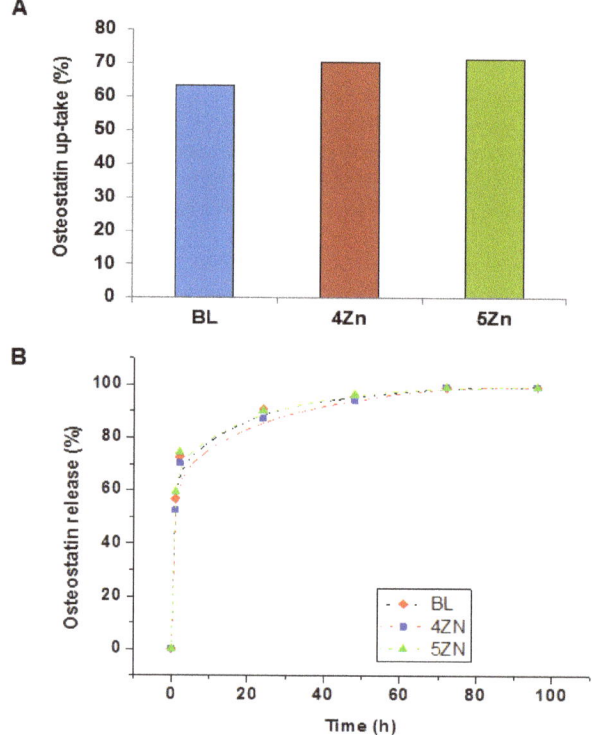

Figure 4. Osteostatin uptake at 24 h (**A**), and its release profiles measured at different times for BL, 4ZN, and 5ZN disks (**B**). Points tracing the curve are the means of three experiments.

This divergence has been dealt with by the introduction of an empirical non-ideality factor δ in Equation (2) [50]:

$$W_t/W_0 = A(1 - \exp^{k_1 \cdot t})^\delta \qquad (2)$$

The values of this non-ideality factor δ range from 1–0 for materials that either follow first-order kinetics or initially release the peptide molecules located on the external surface of the matrices. The obtained data were fitted using this semi-empirical first-order model, and the release parameters are shown in Table 3. According to this model, δ gives an idea of the degree of fidelity of this approximation. In all tested MBG matrices, the δ value was low and similar, indicating that a relatively large percentage of osteostatin molecules released from the external surface of the MBGs. Moreover, the percentage of osteostatin released was maximal, indicating virtually no osteostatin long retention by all these MBGs.

Table 3. Kinetic parameters of osteostatin release from BL, 4ZN, and 5ZN materials. (w_0: initial loaded mass: μg ost/g MBG; A: maximum amount of peptide released; k_1: release rate constant; δ: kinetic non-ideality factor; R: goodness of fit.

	W_0 (g/g)	A (%)	$k_1(*10^3)$ (h^{-1})	δ	R
BL	0.80	99.3 ± 3.3	37.7 ± 3	0.14 ± 0.04	0.996
4ZN	0.90	99.7 ± 3.9	31.5 ± 3	0.13 ± 0.04	0.998
5ZN	0.95	99.6 ± 3.9	31.5 ± 4	0.12 ± 0.07	0.996

3.4. In Vitro Bioactivity Assay

The different MBG disks before and after soaking in SBF for different times were characterized by FTIR. This is a very sensible technique for detecting the formation of amorphous calcium phosphate (ACP) and HCA by evaluating the region of the spectra at around 600 cm^{-1}. The presence of a band in this region is characteristic of ACP, and the split of this band in bands at 560 and 603 cm^{-1} is indicative of phosphate in a crystalline environment like the one in nano-HCA [51].

As observed in Figure 5, the behaviour of BL in SBF was different from that of 4ZN and 5ZN. Thus, for the Zn-free MBG, the band at 600 cm^{-1} was visible at 6 h of incubation, and the bands of HCA were already detected at 1 day. For longer times, like 3 days, the FTIR spectra did not suffer additional changes. However, for 4ZN and 5ZN only the band of ACP was observed after 7 days of treatment and the two bands of HCA were not detected even for soaking days as long as 21 days. These results had already been described for bioactive MBGs to which the ZnO additions impeded the formation of HCA in regular SBF (osteostatin-free) [39]. This result was thought to be a consequence of the initial formation of amorphous calcium zinc phosphate, unable to crystallize as HCA. However, in the present study we demonstrated that, although the impregnation of MBGs with 100 nM osteostatin solution exerted a remarkable effect in the MBGs' behaviour in the presence of cells (see the next sections), osteostatin was not found to affect the in vitro formation of ACP or HCA on the tested glasses in SBF.

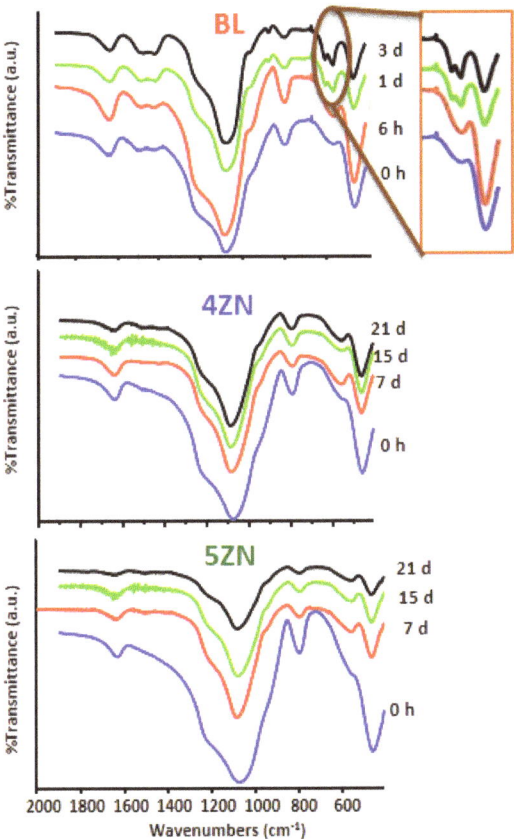

Figure 5. FTIR spectra of BL, 4ZN, and 5ZN after different times in SBF with 100 nM of osteostatin. The ellipse highlights the bands of phosphate in a crystalline environment.

SEM analysis confirmed the FTIR results. Thus, Figure 6 shows the SEM micrographs of BL, 4ZN, and 5ZN before and after being soaked for 7 days in SBF. As is observed, after this time only BL appeared coated by a layer of spherical particles with the characteristic morphology of bone-like HCA. This morphology was developed from the initially formed flocculent shape of ACP. In contrast, for 4ZN and 5ZN disks, no HCA layer was found after 7 days immersed in SBF, although a new material was observed on the 4ZN surface. However, the high intensity of the calcium and phosphorous peaks in the EDX spectra of BL, compared to the other samples, supports the interpretation that HCA preferentially showed up on the BL sample. These results were analogous to those reported for Zn-containing MBGs performed in pure SBF showing the inhibitory effect of Zn^{2+} ions in the HCA crystallization [39]. In the present study, we obtained the identical results demonstrating the null effect of osteostatin additions in the assays in the acellular SBF, in spite of the important effect that exerts in the presence of cells as it will be described in the following sections.

Figure 6. SEM micrographs and EDX spectra of BL, 4ZN, and 5ZN MBG disks, before and after being soaked for 7 days in SBF.

3.5. Degradability of Disks in Complete Medium

To better understand the MBGs cytocompatibility, the release of calcium, phosphorus, and zinc ions from osteostatin-loaded disks after being soaked for 130 h in a complete medium were measured. As it is shown in Figure 7, in BL, calcium and phosphorous concentrations in solution were slightly less than in 4ZN and 5ZN, which can be explained by the HCA layer formed on BL as it was mentioned in the previous section. This variation fit well with the formation of ACP during this interval. Moreover, the Zn concentration in Zn-substituted scaffolds increased until day 5 in both 4ZN and 5ZN disks, reaching a value of 4.6 and 5.3 ppm, respectively; this is consistent with the slightly higher network polymerization of 5ZN material compared to 4ZN, thus releasing less amount of Zn^{2+} to the medium.

3.6. Cell Culture Studies

We next examined and compared the osteogenic activity conferred by Zn^{2+} and osteostatin to these MBGs using MC3T3-E1 pre-osteoblastic cell cultures. We first showed that cell numbers onto the disk's surface at 10 h of culture was increased in both 4ZN and 5ZN materials loaded with osteostatin (Figure 8A). At day 5 of cell culture, the cellular morphology was not modified by any tested material (Figure 8B). Consistent with this result, although Zn^{2+} in these MBGs failed to affect cell number, the presence of osteostatin in both 4ZN and 5ZN materials increased this parameter significantly after 5 days of culture (Figure 9A,B). This pattern of bioactivity matched the amount of Zn^{2+} released to the surrounding medium (Figure 7). Thus, although osteostatin loaded into BL disks exhibited a tendency (but not significant) to increase cell viability (Figure 9A,B), this was only clearly displayed with the peptide-coated 4ZN and 5ZN glasses. None of the tested materials induced significant cell death (about 1%), assessed by Trypan blue exclusion, in these cell cultures (data not shown).

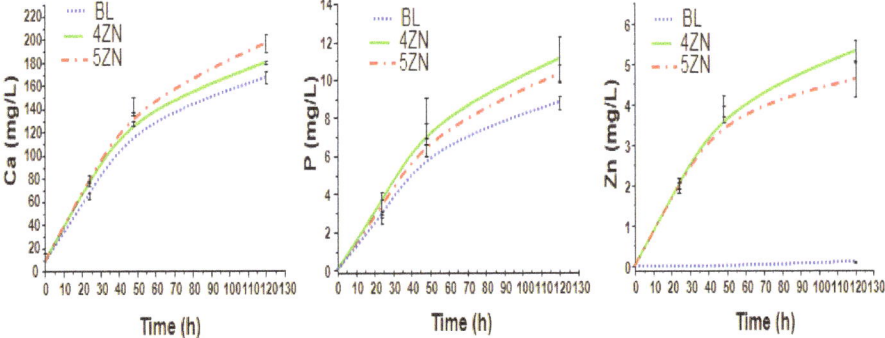

Figure 7. Evolution of cumulative calcium, phosphorus, and zinc content of osteostatin-loaded disks as a function of time in a complete medium.

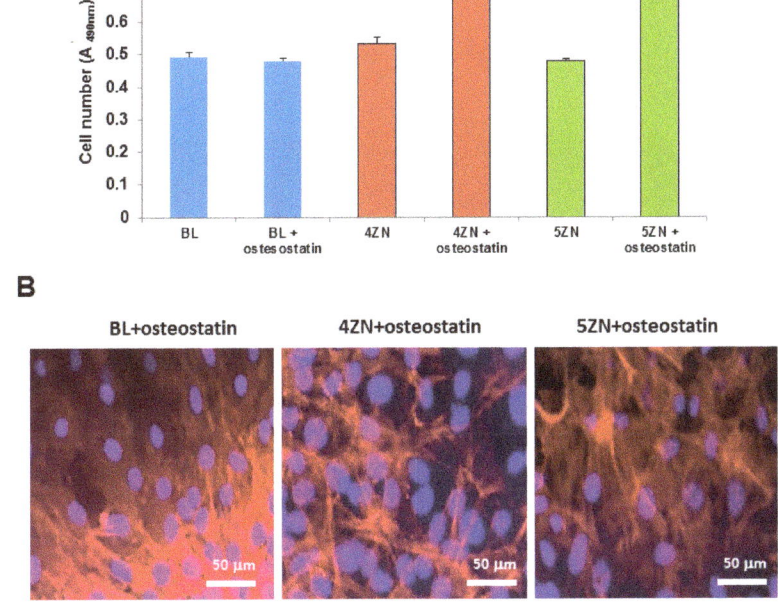

Figure 8. MC3T3-E1 cell number onto BL, 4ZN, and 5ZN disks measured at 10 h of cell culture. (**A**) Results are means ± SEM of three measurements in triplicate (* $p < 0.05$) vs. the corresponding unloaded disks. Absorbance was measured at 490 nm, directly proportional to the number of living adherent cells. Cell morphology evaluation performed by light microscopy onto BL, 4ZN, and 5ZN disks at day 5 of cell culture; (**B**) Cells were stained with DAPI (blue) for the visualization of the cell nuclei and phalloidin-565 (red) for the visualization of cytoplasmic F-actin filaments.

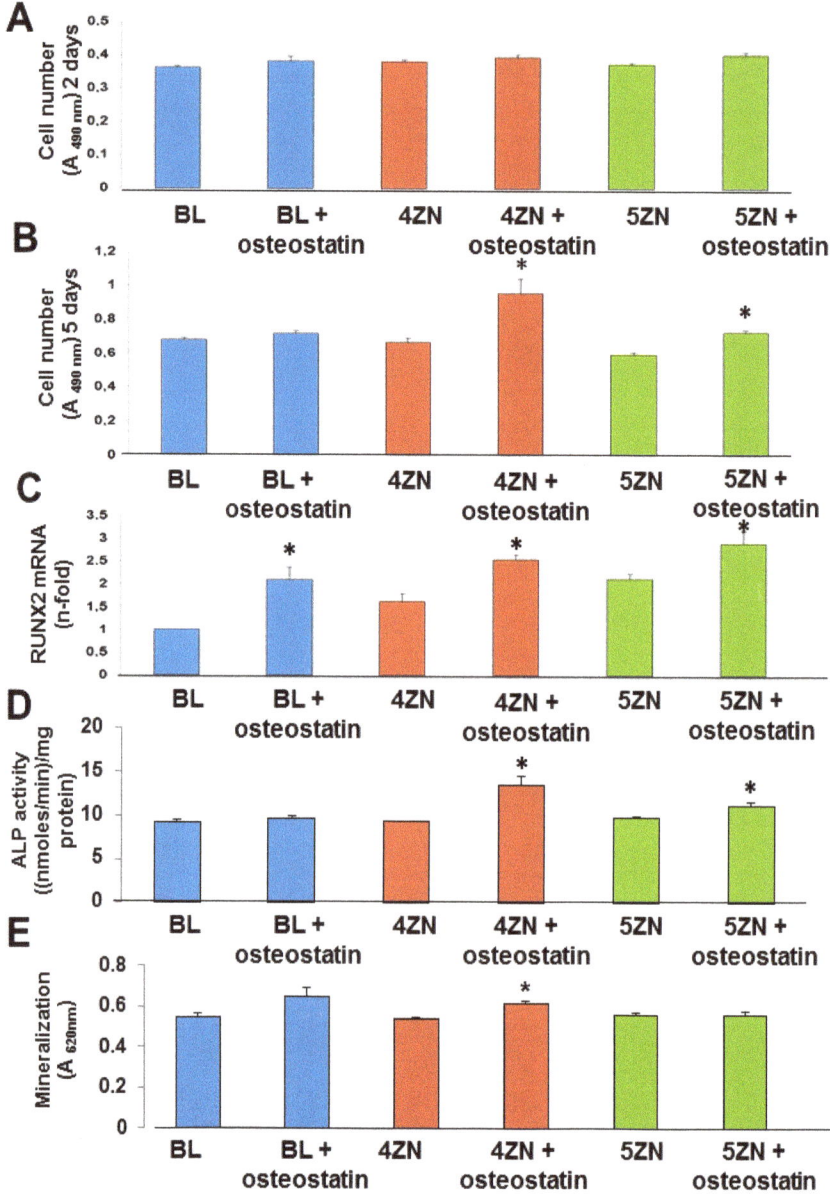

Figure 9. MC3T3-E1 cell number measured by CellTiter 96® AQueous assay in the presence of BL, 4ZN, and 5ZN disks, loaded or not with 100 nM osteostatin, after 2 days (**A**) and 5 days (**B**) of culture. Absorbance was measured at 490 nm, directly proportional to the number of living cells. Runx2 mRNA levels measured by real-time PCR (**C**), ALP activity (**D**), and matrix mineralization measured by Alizarin red staining (**E**) in MC3T3-E1 cells in the presence of these materials at 5 days (**C**,**D**) and 12 days (**E**) of culture. For mineralization studies, absorbance was measured at 620 nm. Results are means ± SEM of three measurements in triplicate (* $p < 0.05$) vs. corresponding unloaded disks.

We next evaluated the capacity of these MBGs to affect osteoblastic cell differentiation. The expression of the early osteoblast differentiation marker Runx2 was increased by the presence of osteostatin in each type of MBG disks at day 5 of MC3T3-E1 cell culture (Figure 9C). Moreover, while 4ZN or 5ZN disks had minimal effect on ALP activity or matrix mineralization in these cells, coating with osteostatin increased these differentiation parameters, mainly in the 4ZN material (Figure 9D,E).

In summary, in vitro studies demonstrated that MBG disks could be loaded with osteostatin, which was mostly released in 24 h. Osteostatin improved the cytocompatibility of Zn-containing MBGs by enhancing osteoblastic proliferation and differentiation without affecting their HCA formation capability. Further studies in vitro and in vivo are needed to elucidate the optimum material, using porous 3-D scaffolds of these MBGs, for bone tissue engineering applications.

4. Conclusions

The results obtained in this study provide a novel and interesting insight in the field of bioactive glasses for bone regeneration. MBG disks containing 4 or 5% ZnO and decorated with osteostatin were shown to improve osteoblastic cell number as well as osteoblast differentiation capacity. For the first time, osteostatin was demonstrated to enhance the in vitro osteogenic capacity of Zn^{2+}-enriched materials, suggesting the potential of this approach in bone tissue engineering applications.

Author Contributions: R.P. and C.H. performed the experiments; S.S.-S. and D.L. supervised the experimental work and wrote the original draft; P.E. and M.V.-R. performed a critical review and revised the publication; A.J.S. organized the financial support, planned the research activity, and supervised the work of the team.

Funding: This research was funded by Instituto de Salud Carlos III, grant number PI15/00978 co-financed with the European Union FEDER funds, the European Research Council, Advanced Grant Verdi-Proposal No. 694160 (ERC-2015-AdG) and Ministerio de Economía y Compatitividad (MINECO), grant number MAT2015-64831-R.

Conflicts of Interest: The authors declare no conflict of interest.

References

1. Salinas, A.J.; Esbrit, P.; Vallet-Regí, M. A tissue engineering approach based on the use of bioceramics for bone repair. *Biomater. Sci.* **2013**, *1*, 40–51. [CrossRef]
2. Wan, C.; Gilbert, S.R.; Wang, Y.; Cao, X.; Shen, X.; Ramaswamy, G.; Jacobsen, K.A.; Alaql, Z.S.; Eberhardt, A.W.; Gerstenfeld, L.C.; et al. Activation of the hypoxia-inducible factor-1α pathway accelerates bone regeneration. *Proc. Natl. Acad. Sci. USA* **2008**, *105*, 686–691. [CrossRef] [PubMed]
3. Yan, X.; Yu, C.; Zhou, X.; Tang, J.; Zhao, D. Highly Ordered Mesoporous Bioactive Glasses with Superior in Vitro Bone-Forming Bioactivities. *Chem. Int. Ed.* **2004**, *43*, 5980–5984. [CrossRef] [PubMed]
4. Vallet-Regí, M.; Salinas, A.J.; Arcos, D. Tailoring the Structure of Bioactive Glasses: From the Nanoscale to Macroporous Scaffolds. *Int. J. Appl. Glass Sci.* **2016**, *7*, 195–205. [CrossRef]
5. Izquierdo-Barba, I.; Arcos, D.; Sakamoto, Y.; Terasaki, O.; López-Noriega, A. High-Performance Mesoporous Bioceramics Mimicking Bone Mineralization. *Chem. Mater.* **2008**, *20*, 3191–3198. [CrossRef]
6. Ito, A.; Kawamura, H.; Otsuka, M.; Ikeuchi, M.; Ohgushi, H.; Ishikawa, K. Zinc-releasing calcium phosphate for stimulating bone formation. *Mater. Sci. Eng. C* **2002**, *22*, 21–25. [CrossRef]
7. Lansdown, A.B.; Mirastschjski, U.; Stubbs, N.; Scanlon, E.; Agren, M.S. Zinc in wound healing: Theoretical, experimental, and clinical aspects. *Wound Repair Regen.* **2007**, *15*, 2–16. [CrossRef] [PubMed]
8. Hoppe, A.; Güldal, N.S.; Boccaccini, A.R. A review of the biological response to ionic dissolution products from bioactive glasses and glass-ceramics. *Biomaterials* **2011**, *32*, 2757–2774. [CrossRef] [PubMed]
9. Zheng, K.; Lu, M.; Rutkowski, B.; Dai, X.; Yang, Y.; Taccardi, N.; Stachewicz, U.; Czyrska-Filemonowicz, A.; Hüser, N.; Boccaccini, A.R. ZnO quantum dots modified bioactive glass nanoparticles with pH-sensitive release of Zn ions, fluorescence, antibacterial and osteogenic properties. *J. Mater. Chem. B* **2016**, *4*, 7936–7949. [CrossRef]
10. Laurenti, M.; Cauda, V. ZnO Nanostructures for Tissue Engineering Applications. *Nanomaterials* **2017**, *7*, 374. [CrossRef] [PubMed]

11. Schierholz, J.M.; Beuth, J. Implant infections: A haven for opportunistic bacteria. *J. Hosp. Infect.* **2001**, *49*, 87–93. [CrossRef] [PubMed]
12. Sánchez-Salcedo, S.; Colilla, M.; Izquierdo-Barba, I.; Vallet-Regí, M. Design and preparation of biocompatible zwitterionic hydroxyapatite. *J. Mater. Chem.* **2013**, *1*, 1595–1606. [CrossRef]
13. Vardakas, K.Z.; Kontopidis, I.; Gkegkes, I.D.; Rafailidis, P.I.; Falagas, M.E. Incidence, characteristics, and outcomes of patients with bone and joint infections due to community-associated methicillin-resistant Staphylococcus aureus: A systematic review. *Eur. J. Clin. Microbiol. Infect. Dis.* **2013**, *32*, 711–721. [CrossRef] [PubMed]
14. Sanchez-Salcedo, S.; Shruti, S.; Salinas, A.J.; Malavasi, G.; Menabue, L.; Vallet-Regí, M. In vitro antibacterial capacity and cytocompatibility of SiO_2–CaO–P_2O_5 meso-macroporous glass scaffolds enriched with ZnO. *J. Mater. Chem. B* **2014**, *2*, 4836–4847. [CrossRef]
15. Lozano, D.; Manzano, M.; Doadrio, J.C.; Salinas, A.J.; Vallet-Regí, M.; Gómez-Barrena, E.; Esbrit, P. Osteostatin-loaded bioceramics stimulate osteoblastic growth and differentiation. *Acta Biomater.* **2010**, *6*, 797–803. [CrossRef] [PubMed]
16. Esbrit, P.; Alcaraz, M.J. Current perspectives on parathyroid hormone (PTH) and PTH-related protein (PTHrP) as bone anabolic therapies. *Biochem. Pharmacol.* **2013**, *85*, 1417–1423. [CrossRef] [PubMed]
17. Datta, N.S.; Abou-Samra, A.B. PTH and PTHrP signalling in osteoblasts. *Cell Signal.* **2009**, *21*, 1245–1254. [CrossRef] [PubMed]
18. Fenton, A.J.; Kemp, B.E.; Hammonds, R.G.; Mitchelhill, K.; Moseley, J.M.; Martin, T.J.; Nicholson, G.C. A potent inhibitor of osteoclastic bone resorption within a highly conserved pentapeptide region of parathyroid hormone-related protein; PTHrP. *Endocrinology* **1991**, *129*, 3424–3426. [CrossRef] [PubMed]
19. Cornish, J.; Callon, K.E.; Nicholson, G.C.; Reid, I.R. Parathyroid hormone-related protein-(107–139) inhibits bone resorption in vivo. *Endocrinology* **1997**, *138*, 1299–1304. [CrossRef] [PubMed]
20. Cornish, J.; Callon, K.E.; Lin, C.; Xiao, C.; Moseley, J.M.; Reid, I.R. Stimulation of osteoblast proliferation by C-terminal fragments of parathyroid hormone-related protein. *J. Bone Mine. Res.* **1999**, *14*, 915–922. [CrossRef] [PubMed]
21. Lozano, D.; De Castro, L.F.; Dapía, S.; Andrade-Zapata, I.; Manzarbeitia, F.; Alvarez-Arroyo, M.V.; Gómez-Barrena, E.; Esbrit, P. Role of Parathyroid Hormone-Related Protein in the Decreased Osteoblast Function in Diabetes-Related Osteopenia. *Endocrinology* **2009**, *150*, 2027–2035. [CrossRef] [PubMed]
22. Lozano, D.; Fernández-de-Castro, L.; Portal-Núñez, S.; López-Herradón, A.; Dapía, S.; Gómez-Barrena, E.; Esbrit, P. The C-terminal fragment of parathyroid hormone-related peptide promotes bone formation in diabetic mice with low-turnover osteopenia. *Br. J. Pharmacol.* **2011**, *162*, 1424–1438. [CrossRef] [PubMed]
23. Rihani-Basharat, S.; Lewinson, D. PTHrP(107–111) Inhibits In Vivo Resorption that was Stimulated by PTHrP(1–34) When Applied Intermittently to Neonatal Mice. *Calcif. Tissue Int.* **1997**, *61*, 426–428. [CrossRef] [PubMed]
24. Fenton, A.J.; Kemp, B.E.; Kent, G.N.; Moseley, J.M.; Zheng, M.H.; Rowe, D.J.; Britto, J.M.; Martin, T.J.; Nicholson, G.C. A Carboxyl-Terminal Peptide from the Parathyroid Hormone-Related Protein Inhibits Bone Resorption by Osteoclasts. *Endocrinology* **1991**, *129*, 1762–1768. [CrossRef] [PubMed]
25. De Gortázar, A.R.; Alonso, V.; Alvarez-Arroyo, M.V.; Esbrit, P. Transient Exposure to PTHrP (107–139) Exerts Anabolic Effects through Vascular Endothelial Growth Factor Receptor 2 in Human Osteoblastic Cells in Vitro. *Calcif. Tissue Int.* **2006**, *79*, 360–369. [CrossRef] [PubMed]
26. Lozano, D.; Trejo, C.G.; Gómez-Barrena, E.; Manzano, M.; Doadrio, J.C.; Salinas, A.J.; Vallet-Regí, M.; García-Honduvilla, N.; Esbrit, P.; Buján, J. Osteostatin-loaded onto mesoporous ceramics improves the early phase of bone regeneration in a rabbit osteopenia model. *Acta Biomater.* **2012**, *8*, 2317–2323. [CrossRef] [PubMed]
27. Trejo, C.G.; Lozano, D.; Manzano, M.; Doadrio, J.C.; Salinas, A.J.; Dapía, S.; Gómez-Barrena, E.; Vallet-Regí, M.; García-Honduvilla, N.; Buján, J.; et al. The osteoinductive properties of mesoporous silicate coated with osteostatin in a rabbit femur cavity defect model. *Biomaterials* **2010**, *31*, 8564–8573. [CrossRef] [PubMed]
28. Lozano, D.; Sánchez-Salcedo, S.; Portal-Nuñez, S.; Vila, M.; López-Herradón, A.; Ardura, J.A.; Mulero, F.; Gomez-Barrena, E.; Vallet-Regi, M.; Esbrit, P. Parathyroid hormone-related protein (107–111) improves the bone regeneration potential of gelatin–glutaraldehyde biopolymer-coated hydroxyapatite. *Acta Biomater.* **2014**, *10*, 3307–3316. [CrossRef] [PubMed]

29. Ardura, J.A.; Portal-Núñez, S.; Lozano, D.; Gutiérrez-Rojas, I.; Sánchez-Salcedo, S.; López-Herradón, A.; Mulero, F.; Villanueva-Peñacarrillo, M.L.; Vallet-Regí, M.; Esbrit, P. Local delivery of parathyroid hormone-related protein-derived peptides coated onto a hydroxyapatite-based implant enhances bone regeneration in old and diabetic rats. *J. Biomed. Mater. Res. A* **2016**, *104*, 2060–2070. [CrossRef] [PubMed]
30. Haimi, S.; Gorianc, G.; Moimas, L.; Lindroos, B.; Huhtala, H.; Raty, S.; Kuokkanen, H.; Sandor, G.K.; Schmid, C.; Miettinen, S.; et al. Characterization of zinc-releasing three-dimensional bioactive glass scaffolds and their effect on human adipose stem cell proliferation and osteogenic differentiation. *Acta Biomater.* **2009**, *5*, 3122–3131. [CrossRef] [PubMed]
31. Salih, V.; Patel, A.; Knowles, J.C. Zinc-containing phosphate-based glasses for tissue engineering. *Biomed. Mater.* **2007**, *2*, 11–20. [CrossRef] [PubMed]
32. Aina, V.; Malavasi, G.; Fiorio, P.A.; Munaron, L.; Morterra, C. Zinc-containing bioactive glasses: Surface reactivity and behaviour towards endothelial cells. *Acta Biomater.* **2009**, *5*, 1211–1222. [CrossRef] [PubMed]
33. Brunauer, S.; Emmet, P.H.; Teller, E. Adsorption of Gases in Multimolecular Layers. *J. Am. Chem. Soc.* **1938**, *60*, 309–319. [CrossRef]
34. Barrett, E.P.; Joyner, L.G.; Halenda, P.P. The Determination of Pore Volume and Area Distributions in Porous Substances. I. Computations from Nitrogen Isotherms. *J. Am. Chem. Soc.* **1951**, *73*, 373–380. [CrossRef]
35. Maçon, A.L.; Kim, T.B.; Valliant, E.M.; Goetschius, K.; Brow, R.K.; Day, D.E.; Hoppe, A.; Boccaccini, A.R.; Kim, I.Y.; Ohtsuki, C.; et al. A unified in vitro evaluation for apatite-forming ability of bioactive glasses and their variants. *J. Mater. Sci. Mater. Med.* **2015**, *26*, 115. [CrossRef] [PubMed]
36. Wang, L.; Nancollas, G.H. Calcium Orthophosphates: Crystallization and Dissolution. *Chem. Rev.* **2008**, *108*, 4628–4669. [CrossRef] [PubMed]
37. Salinas, A.J.; Vallet-Regí, M. Glasses in bone regeneration: A multiscale issue. *J. Non-Cryst. Sol.* **2016**, *432*, 9–14. [CrossRef]
38. Reddi, A.H.; Huggins, C.B. Citrate and alkaline phosphatase during transformation of fibroblasts by the matrix and minerals of bone. *Proc. Soc. Exp. Biol. Med.* **1972**, *140*, 807–810. [CrossRef] [PubMed]
39. Salinas, A.J.; Shruti, S.; Malavasi, G.; Menabue, L.; Vallet-Regí, M. Substitutions of cerium, gallium and zinc in ordered mesoporous bioactive glasses. *Acta Biomater.* **2011**, *7*, 3452–3458. [CrossRef] [PubMed]
40. Shruti, S.; Salinas, A.J.; Malavasi, G.; Lusvardi, G.; Menabue, L.; Ferrara, G.; Mustarelli, P.; Vallet-Regi, M. Structural and in vitro study of cerium, gallium and zinc containing sol–gel bioactive glasses. *J. Mater. Chem.* **2012**, *22*, 13698–13706. [CrossRef]
41. Serra, J.; Gonzalez, P.; Liste, S.; Chiussi, S.; Leon, B.; Perez-Amor, M.; Ylanen, H.O.; Hupa, M. Influence of the non-bridging oxygen groups on the bioactivity of silicate glasses. *J. Mater. Sci. Mater. Med.* **2002**, *13*, 1221–1225. [CrossRef] [PubMed]
42. Turdean-Ionescu, C.; Stevensson, B.; Izquierdo-Barba, I.; García, A.; Arcos, D.; Vallet-Regí, M.; Edén, M. Surface Reactions of Mesoporous Bioactive Glasses Monitored by Solid-State NMR: Concentration Effects in Simulated Body Fluid. *J. Phys. Chem. C* **2016**, *120*, 4961–4974. [CrossRef]
43. Tsai, T.W.T.; Chan, J.C.C. Recent Progress in the Solid-State NMR Studies of Biomineralization. *Ann. Rep. NMR* **2011**, *73*, 1–61.
44. Leonova, E.; Izquierdo-Barba, I.; Arcos, D.; Lopez-Noriega, A.; Hedin, N.; Vallet-Regi, M.; Eden, M. Multinuclear Solid-State NMR Studies of Ordered Mesoporous Bioactive Glasses. *J. Phys. Chem. C* **2008**, *112*, 5552–5562. [CrossRef]
45. Garcıa, A.; Cicuendez, M.; Izquierdo-Barba, I.; Arcos, D.; Vallet-Regí, M. Essential Role of Calcium Phosphate Heterogeneities in 2D-Hexagonal and 3D-Cubic $SiO_2-CaO-P_2O_5$ Mesoporous Bioactive Glasses. *Chem. Mater.* **2009**, *21*, 5474–5484. [CrossRef]
46. Linati, L.; Lusvardi, G.; Malavasi, G.; Menabue, L.; Menziani, M.C.; Mustarell, P.; Segre, U. Qualitative and Quantitative Structure−Property Relationships Analysis of Multicomponent Potential Bioglasses. *J. Phys. Chem. B* **2005**, *109*, 4989–4998. [CrossRef] [PubMed]
47. Izquierdo-Barba, I.; Salinas, A.J.; Vallet-Regí, M. Bioactive glasses: From macro to Nano. *Int. J. Appl. Glass Sci.* **2013**, *4*, 149–161. [CrossRef]
48. Balas, F.; Manzano, M.; Horcajada, P.; Vallet-Regí, M. Confinement and controlled release of bisphosphonates on ordered mesoporous silica-based materials. *J. Am. Chem. Soc.* **2006**, *128*, 8116–8117. [CrossRef] [PubMed]

49. Crank, J. *The Mathematics of Diffusion*, 2nd ed.; Springer: New York, NY, USA, 1975.
50. Manzano, M.; Aina, V.; Arean, C.O.; Balas, F.; Cauda, V.; Colilla, M.; Delgado, M.R.; Vallet-Regi, M. Studies on MCM-41 mesoporous silica for drug delivery: Effect of particle morphology and amine functionalisation. *Chem. Eng. J.* **2008**, *137*, 30–37. [CrossRef]
51. Elliott, J.C. *Structures and Chemistry of the Apatites and Other Calcium Orthophosphates*; Elsevier: Amsterdam, The Netherlands, 1994; Volume 1, pp. 1–201.

© 2018 by the authors. Licensee MDPI, Basel, Switzerland. This article is an open access article distributed under the terms and conditions of the Creative Commons Attribution (CC BY) license (http://creativecommons.org/licenses/by/4.0/).

Article

Mechanical and Physicochemical Properties of Newly Formed ZnO-PMMA Nanocomposites for Denture Bases

Mariusz Cierech [1,*], Izabela Osica [2,3], Adam Kolenda [1], Jacek Wojnarowicz [4], Dariusz Szmigiel [5], Witold Łojkowski [4], Krzysztof Kurzydłowski [2], Katsuhiko Ariga [3] and Elżbieta Mierzwińska-Nastalska [1]

1. Department of Prosthodontics, Medical University of Warsaw, 02-006 Warsaw, Poland; adam.kolenda@poczta.fm (A.K.); elzbieta.mierzwinska-nastalska@wum.edu.pl (E.M.-N.)
2. Faculty of Materials Science and Engineering, Warsaw University of Technology, 02-504 Warsaw, Poland; izabela.osica@gmail.com (I.O.); Krzysztof.kurzydlowski@pw.edu.pl (K.K.)
3. World Premier International Center for Materials Nanoarchitectonics (WPI-MANA), National Institute for Materials Science (NIMS), 1-1 Namiki, Tsukuba 305-0044, Japan; ARIGA.Katsuhiko@nims.go.jp
4. Institute of High Pressure Physics, Polish Academy of Sciences, 01-142 Warsaw, Poland; jacek.wojnarowicz@tlen.pl (J.W.); w.lojkowski@labnano.pl (W.Ł.)
5. Division of Silicon Microsystem and Nanostructure Technology, Institute of Electron Technology, 02-668 Warsaw, Poland; szmigiel@ite.waw.pl
* Correspondence: mariusz.cierech@wp.pl; Tel.: +48-22-502-18-86

Received: 21 March 2018; Accepted: 30 April 2018; Published: 6 May 2018

Abstract: Aim: The aim of this study was to investigate the selected properties of zinc oxide-polymethyl methacrylate (ZnO-PMMA) nanocomposites that can influence the microorganism deposition on their surface. Materials and Methods: Non-commercial ZnO-NPs were prepared, characterized and used for the preparation of PMMA nanocomposite. Roughness, absorbability, contact angle and hardness of this new nanomaterial were evaluated. PMMA without ZnO-NPs served as control. Outcomes: Compared to unenriched PMMA, incorporation of ZnO-NPs to 7.5% for PMMA nanocomposite increases the hardness (by 5.92%) and the hydrophilicity. After modification of the material with zinc oxide nanoparticles the roughness parameter did not change. All tested materials showed absorption within the range of 1.82 to 2.03%, which meets the requirements of International Organization for Standardization (ISO) standards for denture base polymers. Conclusions: The results showed no significant deterioration in the properties of acrylic resin that could disqualify the nanocomposite for clinical use. Increased hydrophilicity and hardness with absorbability within the normal range can explain the reduced microorganism growth on the denture base, as has been proven in a previous study.

Keywords: denture stomatitis; polymethyl methacrylate; zinc oxide nanoparticles; *Candida albicans*

1. Introduction

With the aging of the population, the demand for prosthetic treatment in the stomatognathic system is growing [1]. In toothlessness or multiple teeth deficiencies, removable appliances with an extensive base are still the most common prosthetic solution. The force released in the act of chewing is completely passed by the prosthesis base onto the mucous membrane, the periosteum and bone. This creates the need for taking over by mucosa non-physiological functions, which results in its impaired physiology and increased susceptibility to infection. From the clinical point of view, polymethyl methacrylate (PMMA) has acceptable mechanical properties, nevertheless attempts have still been made to modify the material so that it would become resistant to microbial adhesion.

Some very promising nanoparticles (NPs) that can be used to modify the biomaterial for denture bases are zinc oxide (ZnO) NPs. Zinc oxide is a multi-functional material being a II-VI semiconductor [2]. Properties such as wide band gap (~3.37 eV) or high exciton binding energy (~60 meV) make it an attractive material for electronics and optoelectronics [3,4]. Selective doping of ZnO (NPs, alloys) with ions, for example, Cd^{2+}, Co^{2+}, Mn^{2+} permits changing its optical, magnetic and antibacterial properties, conductivity and photoluminescence [5–8]. The development of methods for obtaining nano-ZnO with a controlled shape and size enabled the commencement of research on the use of ZnO nanomaterials for the production of solar cells, gas sensors, switching memory devices and for water filtration [9]. In industry, ZnO is widely used as an ingredient of rubber, pigments, cements, plastics, sealants and paints. It is a component of pharmaceutical formulations and cosmetics, for example, baby powders, toothpastes and tooth dressings, sunscreens and skin protection balms. The advantages of ZnO-NPs application as a ultraviolet (UV) absorber in the field of personal hygiene and sun protection include long-term protection, broadband protection (UVA and UVB) and non-whitening effect on the skin [9–11]. It is also used in deodorants, medical and sanitary materials, glass, ceramic and self-cleaning materials [10,11]. The antibacterial and antifungal activity of ZnO-NPs was the major rationale for commencing in 2014 the research on dentures modified by ZnO-NPs, which are supposed to ultimately be characterized by antifungal properties [12,13]. At present, this issue has gained a growing interest among the scientist around the world and has already been investigated by several research groups [14–20].

The intention of our research group was to create new biomaterial for denture bases that could impede the adhesion of microorganisms to their surface, thereby decreasing development of denture stomatitis. It is of great importance for patients with immune system deficiencies (e.g., immunotherapy, AIDS, old age) where local fungal inflammation can lead to pneumonia or systemic fungemia endangering the patient's life [21]. To achieve this goal, the modification of polymethyl methacrylate with zinc oxide nanoparticles was performed. The first article in a series [12] about new ZnO-PMMA nanocomposites presented the characteristics of ZnO-NPs (mean particle size, density, specific surface area). The minimal inhibitory concentration for standard strain of *Candida albicans* was determined at the level of 0.75 mg/mL. There, the process of performing nanocomposite was also described and the encountered difficulties presented in uniform distribution of ZnO-NPs in polymer matrix. Nevertheless, the sonication process decreased ZnO-NPs conglomerates, providing more favourable conditions for more homogenic material. Moreover, the SEM taken scans at a magnification of 50 k × confirmed that the material also contained particles smaller than 100 nm meeting the conditions required for nanomaterials. To our best knowledge this was the first successful attempt to produce PMMA resin for bases of dentures modified with nanoparticles of zinc oxide. The second publication [13] discussed the anti-fungal properties of newly formed biomaterial. These studies evidenced the antifungal activity of PMMA-ZnO nanocomposites. The study of the biofilm deposition on the surface showed that antifungal properties increase with increasing concentration of ZnO-NPs. A 7.5% nanocomposite revealed the lowest total amount of *Candida*, a 4.6-fold lower than in PMMA without modification. The (2,3-Bis-(2-Methoxy-4-Nitro-5-Sulfophenyl)-2H-Tetrazolium-5-Carboxanilide)(XTT) assay, in conjunction with testing the turbidity of solutions, may indicate the mechanism by which ZnO-NPs exert their effect on the increased induction of antioxidative stress in microorganism cells. The increased production of reactive oxygen species (ROS) can cause fungicidal action. The aim of this study was to investigate selected properties of nanocomposites, which can influence the microorganism deposition on their surface and better explain the results of the previous publication [13].

The basic property decreasing the deposition of food debris is the smoothness of the surface expressed by its roughness. A smaller expansion of the surface results in a reduced number of natural niches for *C. albicans*, which considerably hampers the formation of the fungal biofilm structure [22]. The hardness of the material does not directly affect the deposition of pathogens but becomes an exponent of the resistance supplement to mechanical damage [23]. It occurs during the act of chewing but also during the performance of denture hygiene especially when the patient, against medical advice, uses a substance with a high abrasion index. This can lead to the microdamage of prostheses and thereby increases the roughness parameter.

Another property that affects the deposition of microorganisms is hydrophobic or hydrophilic nature of the surface. With the increase in hydrophilic property the biofilm formation is reduced [24,25]. It is believed that the hydrophilic material interferes with the initial phase of microbial adhesion based on electrostatic interactions, Van der Waals forces and hydrogen interactions. *Candida* species have a very strong affinity to the surface of the acrylic base due to its specific nature but also because of its hydrophobic properties. In the cell wall of *Candida* there are adhesins, mainly mannoproteins, responsible for adhesion to host cells. Formosa et al. [26] have shown that depending on the conditions fungal cells can produce adhesins on their surface. Then they change the conformation of structures known as nanodomains, exhibiting hydrophobic properties and giving fungi affinity to all hydrophobic, abiotic surfaces, including acrylic resin material.

Absorbability of the material is also its negative feature. In the first stage of denture use the material absorbs fluids from the oral cavity to its interior. This results in an increase in weight of the material, which does not significantly affect the retention and function of the prosthesis but the liquid absorbed into the interior comprises microorganisms present in the saliva making formation of the fungal biofilm possible [27]. However, some authors [28] do not notice the correlation between the sorption of water and ease of accumulation of microorganisms. The PMMA porosity, regarded as one of the major causes of the biofilm formation, seems to be of minor importance. Studies conducted by Osica et al. [29] on the basis of computed tomography revealed an average porosity of PMMA at 0.01%. However, it should be remembered that this parameter varies over time, depending on the length of the prosthesis use and at a later stage it can play a role in the deposition of microorganisms. The process of biofilm formation and the mechanisms responsible for this process are multi-faceted and are the subject of interest to scientists all around the world. In previous articles, we characterized new material [12] and described antifungal properties of ZnO-PMMA nanocomposite [13]. The aim of this study was to investigate selected properties of ZnO-PMMA nanocomposites which can influence the microorganism deposition on their surface.

2. Materials and Methods

2.1. Synthesis and Characteristics of ZnO-NPs

ZnO-NPs were obtained using our microwave solvothermal synthesis method that permits a precise size control of the formed ZnO-NPs [12,13,30–33]. This method is characterized by the repeatability and reproducibility of properties of the synthesised ZnO-NPs. The following characteristics were defined for the obtained nanoparticles: morphology, skeleton density, specific surface area (SSA), phase purity and average particle size. The test procedures and the analysers employed in the characterization of ZnO-NPs used in this study were described extensively in our earlier publications [12,31,33]. The obtained powder had the following parameters: skeleton density 5.24 ± 0.05 g/cm^3, specific surface area 39 ± 1 m^2/g, average particle size (from SSA) 30 ± 0.1, average crystallite size (from Scherrer's formula) 22–25 nm. ZnO-NPs were composed of single spherical crystallites. The ZnO powder contained only the hexagonal phase of ZnO (JCPDS card No. 36-1451 [30]).

2.2. Nanocomposite Preparation

A procedure of ZnO-PMMA nanocomposite preparation was described in a previously published article [12]. The main steps are described below. A thermally polymerized PMMA resin Superacryl Plus (Spofa Dental, Jiczyn, Czech Republic) was used to manufacture the samples. The recommended mixing ratio was 22 g of powder polymer and 10 mL of liquid monomer, which represents a volume ratio of 3:1. The appropriate amount of solvothermal zinc oxide nanopowder was suspended in liquid monomer of PMMA resin. The mixture was shaken in a Vortex VX-200 shaker (Labnet, Edison, NJ, USA) for 10 min and additionally sonicated for 240 s using an Elmasonic S 10/(H) (30 W, Elma Schmidbauer GmbH, Singen, Germany). Then a calculated amount of PMMA resin powder was added, so that the final 2.5%, 5% and 7.5% mass concentration of ZnO-NPs could be obtained. PMMA resin without ZnO-NPs was used as the control group.

In all the studies described below 3 types of nanocomposites (2.5%, 5% and 7.5%) were used and pure PMMA served as the control group.

2.3. Roughness Assay

The surface roughness was examined using a stylus profiler Dektak XT (Bruker, Billerica, MA, USA). Linear and aerial measurements were made. The former measures were taken in a single line and the latter measures in the area of the sample surface. Linear scans, 5 mm long, were taken step by step every 1 mm for both reference and modified samples. Moreover, topography of nanocomposites surface was imaged by colour three-dimensional (3D) laser microscope (VK-9700K, Keyence, Osaka, Japan). The surface area with dimensions of 1416 × 1000 µm was pictured. Examination was performed on 22 samples for each group of material. The data were evaluated for normal distribution using the Kolmogorov-Smirnov assay. Then after checking the homogeneity of variance (Brown-Forsythe assay) one test either Student's t-test for independent samples (1) or Cochran Cox with separate variance estimate test (2) was performed. The level of significance was established at a p-value = 0.05. All data were computed using the Statistica 10.0 program (StatSoft Inc., Tulsa, OK, USA).

2.4. Contact Angle Assay

The hydrophobic/hydrophilic nature of the material before and after modification was determined by the contact angle. The value of the contact angle θ was determined by measuring the angle between the tested flat surface and the tangent created by a drop of liquid bordering with that body. Measurements were made at room temperature using a goniometer type ContactAngle® system (OCA DeltaPhysics, Filderstadt, Germany). As a measuring liquid, distilled water was used. The drop volume measurement was 1 mL, dosing rate was equal to 2 mL/s. Examination was performed on 46 samples for each group of material. The data were evaluated for normal distribution using the Kolmogorov-Smirnov test. Then after checking the homogeneity of variance (Brown-Forsythe and Levene assays) one of the tests: t-Student for independent samples (1) or Cochran Cox with separate variance estimate test (2) was performed. The level of significance was established at a p-value = 0.05. All data were computed using the Statistica 10.0 program (StatSoft Inc., Tulsa, OK, USA).

2.5. Absorbability Assay

The measurement method for the water absorbency was elaborated based on standard procedure described in ISO 62:2008. To measure the water adsorption the samples with diameter of 20 mm × 20 mm × 2 mm were placed in liquid medium in a closed container. The chosen solution to perform this study was Fusayama's artificial saliva, which mimics the oral environment. The Fusayama artificial saliva was prepared with the following composition: 0.2 g NaCl, 0.2 g KCl, 0.453 g $CaCl_2 \cdot 2H_2O$, 0.345 g $NaH_2PO_4 \cdot 2H_2O$, 0.0025 g $Na_2S \cdot 9H_2O$, 0.5 g urea in 1000 mL of deionized water (pH 7). All chemicals were of analytical grade and used without further purification. All solutions were prepared with Milli Q water purified in the Millipore system. The samples were incubated at a human body temperature (37 °C). Subsequently, the samples weight changes were measured at a certain period of time (after 3, 5 and 7 weeks) utilizing a digital balance after slightly drying with a paper towel. Examination was performed on 24 samples for each group of material. The data were evaluated for normal distribution using the Kolmogorov-Smirnov test. Then after checking the homogeneity of variance (Brown-Forsythe and Levene assays) one of the tests: t-Student for independent samples (1) or Cochran Cox with separate variance estimate test (2) was performed. The level of significance was established at a *p*-value = 0.05. All data were computed using the Statistica 12.0 program (StatSoft Inc., Tulsa, OK, USA).

2.6. Hardness Assay

The hardness was examined according to ISO 868:2003 standard using a Shore durometer with scale D which is appropriate for the measurement of semi-rigid and hard plastics materials. The hardened steel rod configured as a needle pin (30° cone) under an applied force penetrates into the material and the depth of penetration is measured on a scale of 0 to 100. A given force of 44.64 N was applied in a consistent manner for a required period of time (15 s). All measurements were performed at room temperature. Examination was performed on 30 samples for each group of material. The data were evaluated for normal distribution using two independent assays, the Kolmogorov-Smirnov and Shapiro-Wilk tests. Then after checking the homogeneity of variance (Brown-Forsythe and Levene assays) one of the tests: *t*-Student for independent samples (1) or Cochran Cox with separate variance estimate test (2) was performed. The level of significance was established at a *p*-value = 0.05. All data were computed using the Statistica 12.0 program (StatSoft Inc., Tulsa, OK, USA).

3. Results and Discussion

3.1. Roughness Assay

The study of the roughness of acrylic samples modified by the addition of ZnO-NPs are illustrated in box plot (Figure 1). The Ra parameter for control group was 3.99 µm with SD = 1.25. The results for 2.5%, 5% and 7.5% nanocomposites were: 3.70 µm (SD = 0.75); 3.46 µm (SD = 0.91) and 3.80 µm (SD = 0.93), respectively. Statistically significant differences were observed neither between 2.5%, 5% and 7.5% nanocomposites nor between nanocomposites and the control group. The three-dimensional surface imaging was obtained by 3D laser scans (Figure 2). The typical surface topography of the nanocomposites and pure PMMA was similar, characterized by large and deep grooves that were formed during the sample manufacture process. Depending on the content of ZnO nanopowder in nanocomposites a slight difference in surface smoothness could be observed (Figure 3), nevertheless it is not noticeable for roughness parameter measured on a micrometre scale.

There have been many studies showing positive correlation between the surface roughness and deposition of pathogenic microorganisms on the material's surface [22,34,35]. The larger the development of the surface and the larger number of natural niches for microorganisms, the greater the potential for the deposition of *C. albicans*. These studies demonstrated neither positive nor negative effect of the addition of zinc oxide nanoparticles on the roughness parameter. The results are in line with the research carried out by Li et al. [36] who also found no statistically significant differences between PMMA and Ag-PMMA nanocomposite. The surface roughness of the samples averaged several micrometres, therefore, it seems unlikely that the additive particles in the nanometre scale could significantly affect the above parameter. Our study shows, however, that even in the same sample, wherein the average roughness is approximately 3 microns, the areas of high surface heterogeneity can be found (Figures 2 and 3). Most likely, this is due to the manufacturing process of the prosthesis and the fact that it reflects of mucosal surface with all anatomic details (palatal folds or small salivary glands). It should be emphasized that the inner surface is not subjected to the polishing process. The prosthesis production procedure itself is not without significance, where the material is polymerized in gypsum form and can penetrate into the micropores of gypsum. Therefore, it is essential to use gypsum of high quality, adequate hardness and mixed in suitable proportions and even foil models to ensure acceptable smoothness of surface retaining all anatomical details. The roughness of the material is one of the key parameters affecting the deposition of microorganisms on the surface of PMMA and after modification of the material with zinc oxide nanoparticles this parameter does not change.

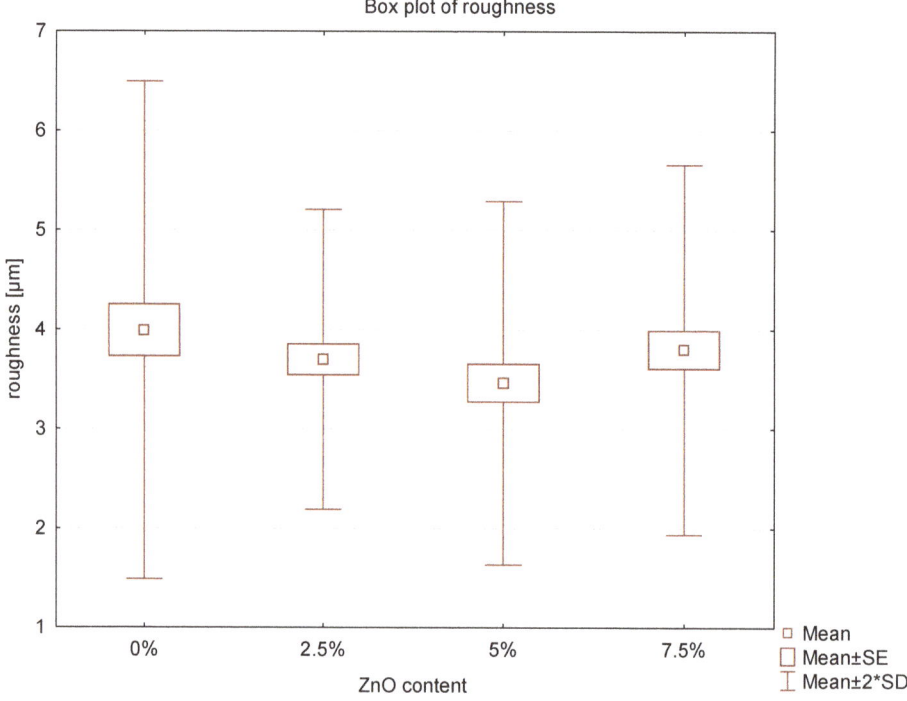

Figure 1. Box plot of the roughness of nanocomposites.

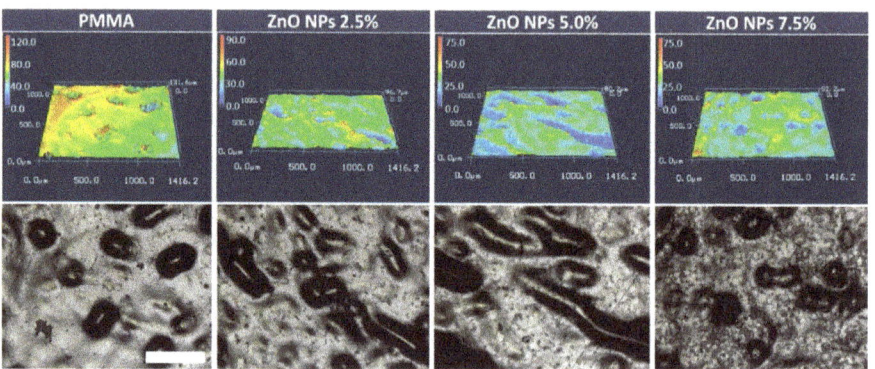

Figure 2. Surface topography of zinc oxide/polymethyl methacrylate (ZnO/PMMA) nanocomposites, the representative 3D laser microscope images (**top**) and optical laser-enhanced images of surfaces (**bottom**). The scale bar (white) refers to 400 µm and it is applicable to all images.

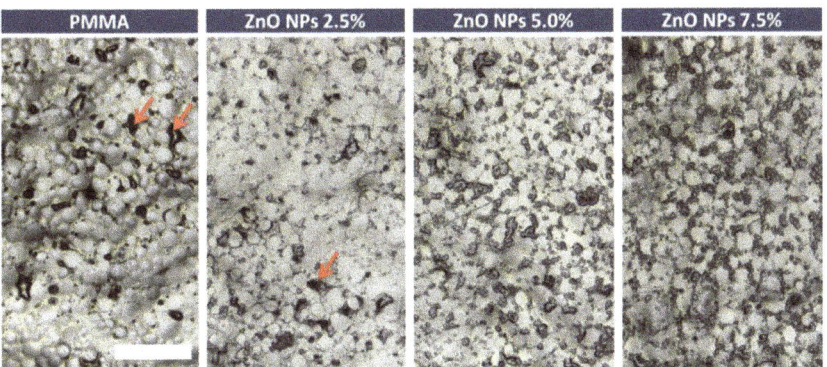

Figure 3. Optical laser-enhanced microscope images of the surfaces of ZnO/PMMA nanocomposites and pure PMMA for comparison. The scale bar (white) refers to 500 μm and it is applicable to all images. The red arrows mark the examples of surface cavities.

3.2. Contact Angle Assay

The results of contact angle measurements are illustrated in a box plot (Figure 4). In all cases of nanocomposites, a statistically significant decrease in the value of the contact angle was observed compared to the control group. The contact angle parameter for the control group was 98.23 with SD = 7.85. Results for 2.5%, 5% and 7.5% nanocomposites were 86.67 (SD = 5.97); 86.36 (SD = 5.42) and 80.97 (SD = 6.03), respectively. No statistically significant differences between the 2.5% and 5% composites were observed, while in 7.5% nanocomposite the decline was at the level of 17.58% compared to the control group. This demonstrates the increased surface hydrophilicity after the addition of ZnO-NPs.

Hydrophobicity or hydrophilicity of materials is assessed on the basis of the contact angle measurement. The larger the angle between the biomaterial surface and liquid drops, the greater the hydrophobic properties exhibited by the biomaterial. The mechanism of microbial adhesion to the prosthesis base consists of two successive phases [37]. One reversible and non-specific uses, among others, electrostatic interaction and van der Waals forces. The other involves specific interactions between the adhesins present in the cell walls of microorganisms and the stereochemically complementar particles on the acrylic plate. *Candida* cells have on their surface hydrophobic polysaccharides, such as mannan or galactomannan, which interact not only with the epithelial mucosa but also with the surface of biomaterials [38]. In further stages of the biofilm formation the interactions between different species of microorganisms begin to play a greater role followed by a rapid growth of *C. albicans*. It has been shown that increasing hydrophilicity of the material hinders the process of microbial deposition [24,25]. Our present study shows that increasing amounts of ZnO-NPs in particular nanocomposites result in an increase in hydrophilic properties of the material. The interaction between water and ZnO nanoparticles leads to both molecular and dissociative adsorption of H_2O forming a number of polar interactions and thus hydrophilic hydroxyl (–OH) species on the surface of particles [39]. The nanoparticles incorporated into the polymer matrix has a significant impact on the surface chemistry of the resulted nanocomposite, increasing its hydrophilicity. Therefore, original hydrophobic interactions between microorganisms and the denture base may be weakened. This finding has clinical implications, because the reduction of the adhesion of the first colonizing microorganisms can disrupt the formation of biofilm already in its initial phase and thus serve as a prophylactic agent for possible inflammation of the prosthetic base.

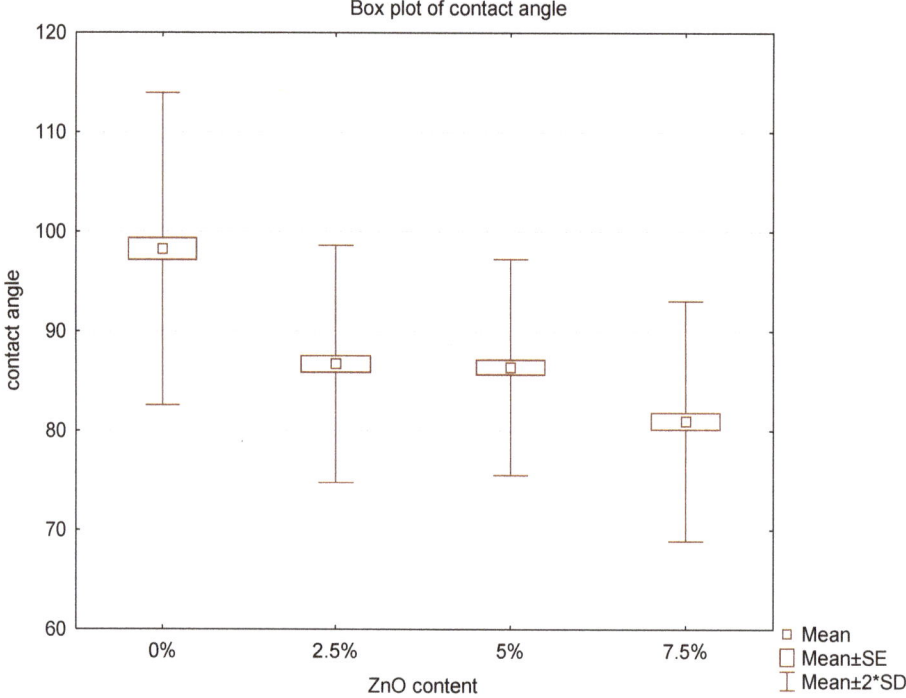

Figure 4. Box plot of the contact angle of nanocomposites.

3.3. Absorbability Assay

The study of the absorbability of acrylic samples modified by the addition of ZnO-NPs after 3 weeks of incubation are illustrated in the box plot (Figure 5). No statistically significant differences were observed between the 2.5% or 5% nanocomposite and the control group, however the examination revealed statistically significant differences between the 7.5% nanocomposite compared to pure PMMA. All the results were within the range of 1.82–2.03%, which meets the requirements of ISO 20795-1:2013 Dentistry—Base polymers—Part 1: Denture base polymers norm. The lowest result was obtained in 7.5% nanocomposite and it was over 10% lower compared to control. Changes in absorption of examined materials, observed in two consecutive series, were small and after 7 weeks of incubation amounted from 0.13% for 2.5% nanocomposite to 0.01% for 5% nanocomposite (Figure 6). Finally, after 7 weeks the absorbability for the control group was 2.05% with SD = 0.08. The results for 2.5%, 5% and 7.5% nanocomposites were: 2.18% (SD = 0.07); 2.07 (SD = 0.09) and 1.90 (SD = 0.15), respectively.

Absorbability of materials specifies their possible absorption of liquids with suspended organic and inorganic components. In the oral environment, these components may include food debris and micro-organisms present in the saliva biocenosis. The higher the absorption, the higher the risk of colonization of biomaterial structures by pathogenic microorganisms and also the higher the probability of faster and less favourable changes in the colour and fragrance of the test material [27]. Water absorption by weight for acrylic resin used as denture bases averages 2% [40]. All tested materials showed absorption within the range of 1.82 to 2.03%, which meets the requirements of ISO 20795-1,2013 Dentistry—Base polymers—Part 1, Denture base polymers. The highest water sorption occurs during the first 3 weeks of incubation, update every 4 months was a maximum of 0.13%. The lowest water absorption was recorded for 7.5% nanocomposite (10% lower compared to the pure PMMA), which may contribute to the reduced colonization of microbes. Therefore, it could be expected that the base plate made of the above-mentioned biomaterial might be able to survive longer in the mouth, without changes in colour and smell.

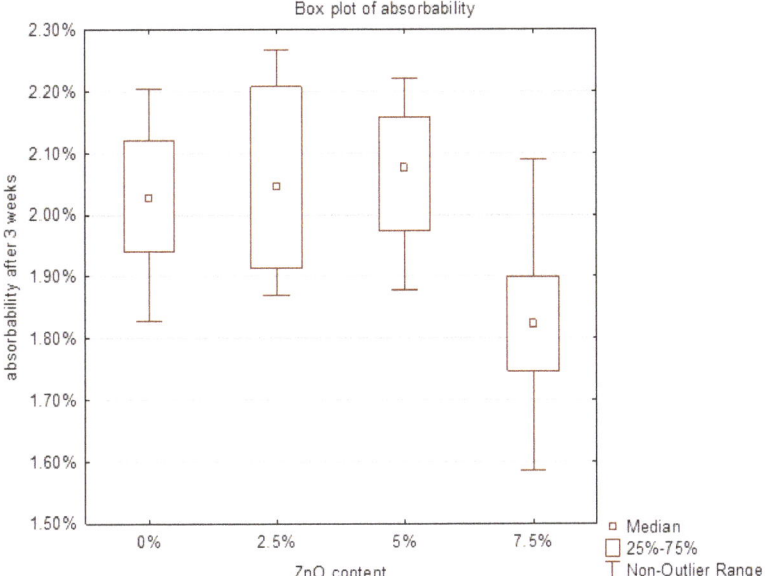

Figure 5. Box plot of the absorbability of nanocomposites after 3 weeks.

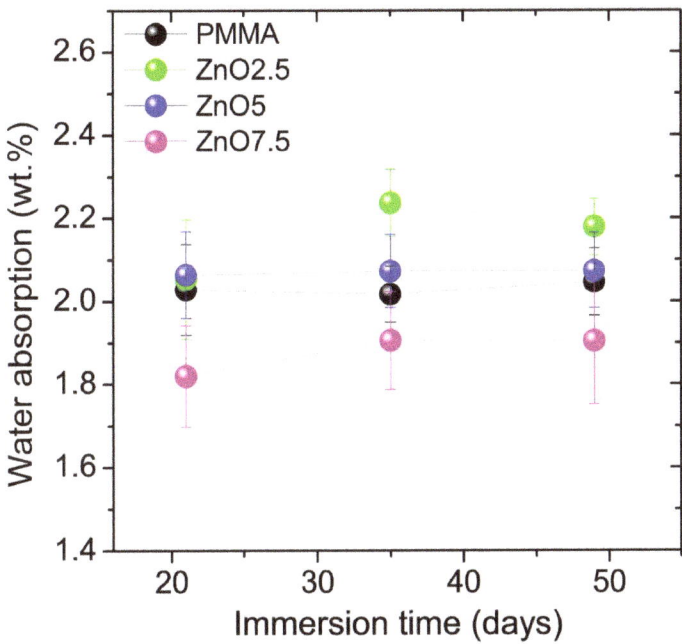

Figure 6. Changes in absorbability of materials after 3, 5 and 7 weeks.

3.4. Hardness Assay

The study of the hardness of acrylic samples modified by the addition of ZnO-NPs are illustrated in the box plot (Figure 7). The determined hardness for the control group was 86.1° ShD with SD = 1.92. The results for 2.5%, 5% and 7.5% nanocomposites were 86.2° ShD (SD = 2.29); 88.7° ShD (SD = 1.31) and 91.2° ShD (SD = 1.56), respectively. No statistically significant differences were observed between the 2.5% nanocomposite and the control group, however the examination revealed statistically significant differences between 5% and 7.5% nanocomposites compared to pure PMMA. Moreover, the highest result was obtained for 7.5% nanocomposite and it was over 5.9% higher compared to control.

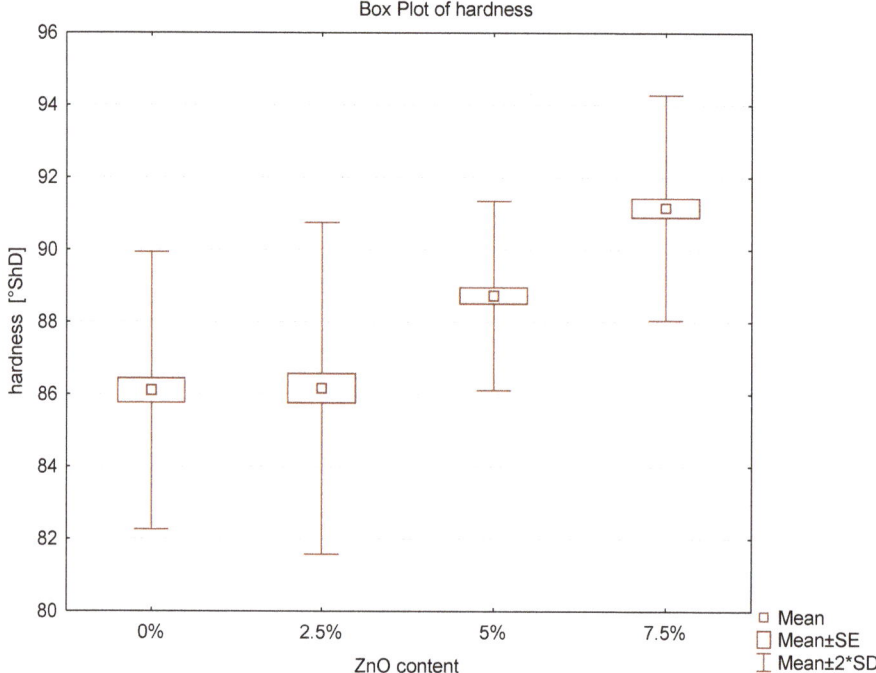

Figure 7. Box plot of the hardness of nanocomposites.

The hardness of the material is not directly linked with the deposition of fungal biofilm on the nanocomposite surface. Nevertheless, a possible use of this biomaterial as a dental prosthesis and the prevailing conditions in the oral cavity in which it will be used should be taken into account. It is essential to subject the denture to hygiene treatment after each meal to initiate the effect of mechanical and chemical stimuli. Patients, often against the doctor's advice, use for denture cleaning a toothpaste with a high abrasion index containing unfavourable for PMMA abrasives that may scratch the surface of the denture. Any damage to denture, such as micro-scratches and cracks, caused during hygiene procedures or during chewing, may be natural niches for living microbes, including the most frequently isolated *C. albicans* in denture stomatitis. The hardness of the material can thus be the exponent of the prosthesis wear resistance capability [41]. According to theoretical considerations the addition of inorganic nanofiller particles to the polymer matrix increases the hardness of the material [42]. The results of this study show that the incorporation of ZnO-NPs to PMMA for 7.5% nanocomposite increases the hardness by 5.92% and this is a statistically significant difference. The revealed increase may hinder the formation of scratches and microcracks on the surface of the prosthesis which explains the reduced deposition of fungal biofilm on the surface of new biomaterials demonstrated in previous publications.

Moreover, the measurement of polymer hardness has been used successfully as an indirect method to assess the degree of conversion of the polymer in acrylic materials conventionally hot polymerized, as well as in composites [43,44]. The parameter of the material hardness is therefore sensitive to the increased content of residual monomer [45], one of the causes of allergic reactions and inflammation of mucosa under the denture base. Based on the presented results it can be concluded that in the entire study group the addition of ZnO-NPs to the polymer matrix does not contribute to the reduced hardness of plastic and thereby to the deteriorated degree of the polymer conversion. This observation is of extreme importance since the increased contents of residual monomer could influence not only the tendency to develop inflammation of the mucous membranes but also to exert a negative effect on mechanical properties of a new biomaterial.

4. Conclusions

This study is a continuation and extension of a previously published work [13] where anti-fungal properties of ZnO-PMMA nanocomposites, were demonstrated and the fact that they increase with increasing content of these zinc oxide nanoparticles was highlighted. The aim of the work was to investigate some properties of new biomaterials that could explain the origin of these properties. The obtained results of laboratory tests are encouraging. However, they must be clinically validated. It would be now interesting to carry out research to find out whether or not the addition of ZnO-NPs contributes at the same time to deterioration of the material mechanical properties. Therefore, prior to clinical application of the new biomaterial further mechanical and cytotoxic studies are required. Finally, it should be noted that the study showed no significant deterioration in the properties of acrylic resin, which could disqualify nanocomposite for clinical use. The increased hydrophilicity and hardness with absorbability within the normal range can explain the reduced microorganisms' growth on the denture base what has been proven in a previous work. The presented study brings much closer the possibility of clinical use of new nanobiomaterials.

Author Contributions: M.C., I.O., A.K., J.W. and D.S. conceived and designed the experiments; M.C., I.O., A.K., J.W. and D.S. performed the experiments; M.C. analyzed the data; M.C. and J.W. contributed reagents and materials; M.C., I.O., A.K., J.W. and D.S. wrote the manuscript; W.Ł., K.K., K.A. and E.M.-N. provided support and critically reviewed the manuscript.

Acknowledgments: The work was supported by a research project carried out in the years 2017–2018, funded by a statutory grant obtained by the Faculty of Medicine and Dentistry, Medical University of Warsaw. The paper was partially prepared as a result of execution of "PRELUDIUM 10" research project, ref. no. UMO-2015/19/N/ST5/03668, financed by the National Science Centre, Poland. Part of the research project was carried out with the use of equipment funded by the project CePT, reference: POIG.02.02.00-14-024/08, financed by the European Regional Development Fund within the Operational Programme "Innovative Economy" for 2007–2013.

Conflicts of Interest: The authors declare no conflict of interest.

References

1. Konopka, T.; Dembowska, E.; Pietruska, M.; Dymalski, P.; Górska, R. Periodontal status and selected indexes of oral cavity condition in Poles aged 65 to 74 years. *Epidemiol. Rev.* **2015**, *69*, 643–647.
2. Fortunato, E.; Gonçalves, A.; Pimentel, A.; Barquinha, P.; Gonçalves, G.; Pereira, L.; Ferreira, I.; Martins, R. Zinc oxide, a multifunctional material: From material to device applications. *Appl. Phys.* **2009**, *96*, 197–205. [CrossRef]
3. Klingshirn, C.F.; Waag, A.; Hoffmann, A.; Geurts, J. *Zinc Oxide*, 1st ed.; Springer: Berlin, Germany, 2010; ISBN 978-3-642-10576-0.
4. Ozgur, U.; Hofstetter, D.; Morkoc, H. 2010 ZnO devices and applications: A review of current status and future prospects. *Proc. IEEE* **2010**, *98*, 1255–1268. [CrossRef]

5. Geetha, N.; Sivaranjani, S.; Ayeshamariam, A.; Suthan Kissinger, J.; Valan Arasu, M.; Jayachandran, M. ZnO doped Oxide Materials: Mini Review. *Fluid Mech. Open Acc.* **2016**, *3*, 141. [CrossRef]
6. Mohanta, A.; Thareja, R.K. Photoluminescence study of ZnCdO alloy. *J. Appl. Phys.* **2008**, *103*, 024901. [CrossRef]
7. Dietl, T.; Ohno, H.; Matsukura, M.; Cibert, J.; Ferrand, D. Zener Model Description of Ferromagnetism in Zinc-Blende Magnetic Semiconductors. *Science* **2000**, *287*, 1019–2000. [CrossRef] [PubMed]
8. Wojnarowicz, J.; Mukhovskyi, R.; Pietrzykowska, E.; Kusnieruk, S.; Mizeracki, J.; Lojkowski, W. Microwave solvothermal synthesis and characterization of manganese-doped ZnO nanoparticles. *Beilstein J. Nanotechnol.* **2016**, *7*, 721–732. [CrossRef] [PubMed]
9. Future Markets, Inc. *The Global Market for Zinc Oxide Nanoparticles*; ID: 3833830; Future Markets, Inc.: Edinburgh, UK, 2016.
10. Future Markets, Inc. *The Global Market for Nanotechnology and Nanomaterials in Cosmetics, Personal Care and Sunscreens*; ID: 3784908; Future Markets, Inc.: Edinburgh, UK, 2016.
11. Zhang, Y.; Nayak, T.R.; Hong, H.; Cai, W. Biomedical Applications of Zinc Oxide Nanomaterials. *Curr. Mol. Med.* **2013**, *13*, 1633–1645. [CrossRef] [PubMed]
12. Cierech, M.; Wojnarowicz, J.; Szmigiel, D.; Bączkowski, B.; Grudniak, A.; Wolska, I.; Łojkowski, W.; Mierzwińska-Nastalska, E. Preparation and characterization of ZnO-PMMA resin nanocomposites for denture bases. *Acta Bioeng. Biomech.* **2016**, *18*, 31–41. [PubMed]
13. Cierech, M.; Kolenda, A.; Grudniak, A.; Wojnarowicz, J.; Woźniak, B.; Gołaś, M.; Swoboda-Kopeć, E.; Łojkowski, W.; Mierzwińska-Nastalska, E. Significance of polymethylmethacrylate (PMMA) modification by zinc oxide nanoparticles for funFagal biofilm formation. *Int. J. Pharm.* **2016**, *510*, 323–335. [CrossRef] [PubMed]
14. Robati Anaraki, M.; Jangjoo, A.; Alimoradi, F.; Maleki Dizaj, S. Comparison of Antifungal Properties of Acrylic Resin Reinforced with ZnO and Ag Nanoparticles. *Pharm. Sci.* **2017**, *23*, 207–214. [CrossRef]
15. Anwander, M.; Rosentritt, M.; Schneider-Feyrer, S.; Hahnel, S. Biofilm formation on denture base resin including ZnO, CaO and TiO_2 nanoparticles. *J. Adv. Prosthodont.* **2017**, *9*, 482–485. [CrossRef] [PubMed]
16. Chen, R.; Han, Z.; Huang, Z.; Karki, J.; Wang, C.; Zhu, B.; Zhang, X. Antibacterial activity, cytotoxicity and mechanical behavior of nano-enhanced denture base resin with different kinds of inorganic antibacterial agents. *Dent. Mater. J.* **2017**, *36*, 693–699. [CrossRef] [PubMed]
17. Gad, M.M.; Fouda, S.M.; Al-Harbi, F.A.; Näpänkangas, R.; Raustia, A. PMMA denture base material enhancement: A review of fiber, filler and nanofiller addition. *Int. J. Nanomed.* **2017**, *12*, 3801–3812. [CrossRef] [PubMed]
18. Kamonkhantikul, K.; Arksornnukit, M.; Takahashi, H. Antifungal, optical and mechanical properties of polymethylmethacrylate material incorporated with silanized zinc oxide nanoparticles. *Int. J. Nanomed.* **2017**, *12*, 2353–2360. [CrossRef] [PubMed]
19. Popovic, P.; Bobovnik, R.; Bolka, S.; Vukadinovic, M.; Lazic, V.; Rudolf, R. Synthesis of PMMA/ZnO nanoparticles composite used for resin teeth. *Mater. Technol.* **2017**, *51*, 871–878. [CrossRef]
20. Salahuddin, N.; El-Kemary, M.; Ibrahim, E. Reinforcement of polymethyl methacrylate denture base resin with ZnO nanostructures. *Int. J. Appl. Ceram. Technol.* **2018**, *15*, 448–459. [CrossRef]
21. Iinuma, T.; Arai, Y.; Abe, Y. Denture Wearing during Sleep Doubles the Risk of Pneumonia in the Very Elderly. *J. Dent. Res.* **2015**, *94* (Suppl. 3), 28S–36S. [CrossRef] [PubMed]
22. Cierech, M.; Szczypińska, A.; Wróbel, K.; Gołaś, M.; Walke, W.; Pochrząst, M.; Przybyłowska, D.; Bielas, W. Comparative analysis of the roughness of acrylic resin materials used in the process of making the bases of dentures and investigation of adhesion of *Candida albicans* to them. *Dent. Med. Probl.* **2013**, *50*, 341–347.
23. Azevedo, A.; Machado, A.L.; Vergani, C.E.; Giampaolo, E.T.; Pavarina, A.C. Hardness of denture base and hard chair-side reline acrylic resins. *J. Appl. Oral Sci.* **2005**, 291–295. [CrossRef]
24. Lazarin, A.; Machado, A.; Zamperini, C.; Wady, A.; Spolidorio, D.; Vergani, C. Effect of experimental photopolymerized coatings on the hydrophobicity of a denture base acrylic resin and on *Candida albicans* adhesion. *Arch. Oral Biol.* **2013**, *58*, 1–9. [CrossRef] [PubMed]
25. Ali, A.A.; Alharbi, F.A.; Suresh, C.S. Effectiveness of coating acrylic resin dentures on preventing *Candida* adhesion. *J. Prosthodont.* **2013**, *22*, 445–450. [CrossRef] [PubMed]

26. Formosa, C.; Schiavone, M.; Boisrame, A.; Richard, M.L.; Duval, R.E.; Dague, E. Multiparametric imaging of adhesive nanodomains at the surface of Candida albicans by atomic force microscopy. *Nanomed. Nanotechnol. Biol. Med.* **2015**, *11*, 57–65. [CrossRef] [PubMed]
27. Kucharski, Z. Physical properties of resilient materials in prosthodontics. *Prosthodontics* **2008**, *63*, 209–216.
28. Gyo, M.; Nikaido, T.; Okada, K.; Yamauchi, J.; Tagami, J.; Matin, K. Surface response of fluorine polymer-incorporated resin composites to cariogenic biofilm adherence. *Appl. Environ. Microbiol.* **2008**, *74*, 1428–1435. [CrossRef] [PubMed]
29. Osica, I.; Cierech, M.; Szlązak, K.; Kapłan, T.; Wróbel, K.; Mierzwińska-Nastalska, E.; Bucki, J.; Święszkowski, W. Investigation of porosity and absorbablity of acrylic resins applied in dental prosthodontics in order to perform bases of dentures. *Przetw. Tworzyw.* **2013**, *151*, 32–36.
30. Wojnarowicz, J.; Opalinska, A.; Chudoba, T.; Gierlotka, S.; Mukhovskyi, R.; Pietrzykowska, E.; Sobczak, K.; Lojkowski, W. Effect of water content in ethylene glycol solvent on the size of ZnO nanoparticles prepared using microwave solvothermal synthesis. *J. Nanomater.* **2016**, *2016*, 2789871. [CrossRef]
31. Wojnarowicz, J.; Chudoba, T.; Koltsov, I.; Gierlotka, S.; Dworakowska, S.; Lojkowski, W. Size control mechanism of ZnO nanoparticles obtained in microwave solvothermal synthesis. *Nanotechnology* **2018**, *29*, 065601. [CrossRef] [PubMed]
32. Majcher, A.; Wiejak, J.; Przybylski, J.; Chudoba, T.; Wojnarowicz, J. A Novel Reactor for Microwave Hydrothermal Scale-up Nanopowder Synthesis. *Int. J. Chem. React. Eng.* **2013**, *11*, 361–368. [CrossRef]
33. Wojnarowicz, J.; Chudoba, T.; Gierlotka, S.; Sobczak, K.; Lojkowski, W. Size Control of Cobalt-Doped ZnO Nanoparticles Obtained in Microwave Solvothermal Synthesis. *Crystals* **2018**, *8*, 179. [CrossRef]
34. Morgan, T.D.; Wilson, M. The effects of surface roughness and type of denture acrylic on biofilm formation by Streptococcus oralis in a constant depth film fermentor. *J. Appl. Microb.* **2001**, *91*, 47–53. [CrossRef]
35. Da Silva, W.J.; Gonçalves, L.M.; Viu, F.C.; Brasil Leal, C.M.; Barbosa, C.M.R.; Del Bel Cury, A.A. Surface roughness influences *Candida albicans* biofilm formation on denture materials. *Rev. Odonto Ciênc.* **2016**, *31*, 54–58. [CrossRef]
36. Li, Z.; Sun, J.; Lan, J.; Qi, Q. Effect of a denture base acrylic resin containing silver nanoparticles on *Candida albicans* adhesion and biofilm formation. *Gerodontology* **2016**, *33*, 209–216. [CrossRef] [PubMed]
37. Waters, M.G.J.; Williams, D.W.; Jagger, R.G.; Lewis, M.A.O. Adherence of *Candida albicans* to experimental denture soft lining materials. *J. Prosthet. Dent.* **1997**, *77*, 306–312. [CrossRef]
38. Calderone, R.A. Molecular interactions at the interface of Candida albicans and host cells. *Arch. Med. Res.* **1993**, *24*, 275–279. [PubMed]
39. Noei, H.; Qiu, H.; Wang, Y.; Loffler, E.; Woll, C.; Muhler, M. The identification of hydroxyl groups on ZnO nanoparticles by infrared spectroscopy. *Phys. Chem. Chem. Phys.* **2008**, *10*, 7092–7097. [CrossRef] [PubMed]
40. Wagner, L. *Introduction to Clinical Practice of Material Science*; Medical University of Warsaw Publishing: Warsaw, Poland, 2007.
41. Craig, R.G.; Powers, J.M. *Restorative Dental Materials*, 11th ed.; Mosby: St. Louis, MO, USA; London, UK, 2002; ISBN 0323014429.
42. Oh, I.S.; Park, N.H.; Suh, K.D. Mechanical and surface hardness properties of ultraviolet-cured polyurethane acrylate anionomer/silica composite film. *J. Appl. Polym. Sci.* **2000**, *75*, 968–975. [CrossRef]
43. Dunn, W.J.; Bush, A.C. A comparison of polymerization by lightemitting diode and halogen-based light-curing units. *J. Am. Dent. Assoc.* **2002**, *133*, 335–341. [CrossRef] [PubMed]
44. Rueggeberg, F.A.; Craig, R.G. Correlation of parameters used to estimate monomer conversion in a light-cured composite. *J. Dent. Res.* **1998**, *67*, 932–937. [CrossRef] [PubMed]
45. Frangou, M.J.; Polyzois, G.L. Effect of microwave polymerization on indentation creep, recovery and hardness of acrylic denture base materials. *Eur. J. Prosthodont. Restor. Dent.* **1993**, *1*, 111–115. [PubMed]

© 2018 by the authors. Licensee MDPI, Basel, Switzerland. This article is an open access article distributed under the terms and conditions of the Creative Commons Attribution (CC BY) license (http://creativecommons.org/licenses/by/4.0/).

Article

Nanostructured ZnO as Multifunctional Carrier for a Green Antibacterial Drug Delivery System—A Feasibility Study

Federica Leone [1], Roberta Cataldo [1], Sara S. Y. Mohamed [1], Luigi Manna [1], Mauro Banchero [1], Silvia Ronchetti [1], Narcisa Mandras [2], Vivian Tullio [2,†], Roberta Cavalli [3,†] and Barbara Onida [1,*]

1. Politecnico di Torino, Department of Applied Science and Technology, Corso Duca Degli Abruzzi 24, 10129 Turin, Italy; federica.leone@polito.it (F.L.); roberta.cataldo@studenti.polito.it (R.C.); sara.mohamed@polito.it (S.S.Y.M.) luigi.manna@polito.it (L.M.); mauro.banchero@polito.it (M.B.); silvia.ronchetti@polito.it (S.R.)
2. Department of Public Health and Pediatrics, Microbiology Division, University of Turin, via Santena 9, 10126 Turin, Italy; narcisa.mandras@unito.it (N.M.); vivian.tullio@unito.it (V.T.)
3. Department of Drug Science and Technology, University of Turin, via Pietro Giuria 9, 10125 Turin, Italy; roberta.cavalli@unito.it
* Correspondence: barbara.onida@polito.it; Tel.: +39-011-090-4631
† These authors contributed equally to this work.

Received: 14 February 2019; Accepted: 2 March 2019; Published: 11 March 2019

Abstract: The physico–chemical and biological properties of nanostructured ZnO are combined with the non-toxic and eco-friendly features of the $scCO_2$-mediated drug loading technique to develop a multifunctional antimicrobial drug delivery system for potential applications in wound healing. Two nanostructured ZnO (NsZnO) with different morphologies were prepared through wet organic-solvent-free processes and characterized by means of powder X-ray diffraction, field emission scanning electron microscopy (FESEM), and nitrogen adsorption analysis. The antimicrobial activity of the two samples against different microbial strains was investigated together with the in vitro Zn^{2+} release. The results indicated that the two ZnO nanostructures exhibited the following activity: *S. aureus* > *C. albicans* > *K. pneumoniae*. A correlation between the antimicrobial activity, the physico–chemical properties (specific surface area and crystal size) and the Zn^{2+} ion release was found. Ibuprofen was, for the first time, loaded on the NsZnO carriers with a supercritical CO_2-mediated drug impregnation process and in vitro dissolution studies of the loaded drug were performed. A successful loading up to 14% w/w of ibuprofen in its amorphous form was obtained. A preliminary drug release test showed that up to 68% of the loaded ibuprofen could be delivered to a biological medium, confirming the feasibility of using NsZnO as a multifunctional antimicrobial drug carrier.

Keywords: Supercritical CO_2; ibuprofen; NsZnO; antimicrobial activity

1. Introduction

Zinc oxide (ZnO) is a multifunctional material possessing unique physical and chemical properties, such as high chemical stability, a high electrochemical coupling coefficient, a broad range of radiation absorption and high photostability [1]. For these reasons it is largely used in many applications, ranging from electronics, optoelectronics, sensors and photocatalysis [2,3].

ZnO is also widely used in topical formulations to address several skin conditions, like burns, scars, and irritations, thanks to its non-toxicity, biocompatibility and antimicrobial activity [4].

ZnO exhibits three crystal structures named wurtzite, zinc-blend, and an occasionally noticed rock-salt [5], which allow it to be employed as a nanostructured material for different nanotechnology applications in many industrial areas, such as gas sensors, biosensors, semiconductors, piezoelectric devices, field emission displays and photocatalytic degradation of pollutants [6].

Due to the wide spread of nanotechnology, cosmetics and pharmaceuticals have also been revolutionized. Among all the materials, ZnO has been developed in different nanostructures to enhance its interaction with the skin and to improve the existing products. A promising application consists of the addition of ZnO in wound dressing materials. Nanocomposites represent a good example [7]: they consist of the addition of ZnO nanostructures to polymeric matrices [8–11] in order to impart novel functionalities, such as antibacterial activity. It is well-established in the literature that ZnO displays significant bactericidal properties over a broad range of bacteria [12,13]. This occurs due to several mechanisms, such as generation of reactive oxygen species (ROS), zinc ion release, membrane dysfunction, and nanoparticle penetration. Moreover the physico–chemical parameters of the nanomaterial, such as size, morphology, and specific surface area, remarkably affect the antibacterial properties of ZnO [14,15]. It has also been demonstrated that zinc ion release can improve wound healing [16,17], since zinc is an essential trace element in the human body and acts as a cofactor in zinc-dependent matrix metalloproteinases that augment auto debridement and keratinocyte migration during wound repair.

The treatment of painful wounds is another important issue in biomedicine. It has been demonstrated that painful wounds can take more time to heal, leading to a lack of compliance by the patients. Several research works [18] have addressed the development of innovative wound dressings, able to deliver small doses of anti-inflammatory analgesic drugs to the wound [19]. In this context the use of ZnO as a drug carrier to be included in the wound dressing device could be of particular interest thanks to its outstanding biocompatibility, even though the application of this material as a drug delivery system has not been investigated widely in the literature [20] and may be considered at its nascent stage.

The use of organic solvents in pharmaceutical technologies is another challenging issue since it leads both to health concerns, which are related to the toxicity of residual solvents in the final products, and to a negative environmental impact. In the last decade, supercritical fluid technology has been emerging as a green drug impregnation method [21]. Supercritical carbon dioxide ($scCO_2$) is the most used supercritical solvent because it is readily available, cheap, non-toxic, non-flammable, and recyclable. At the end of the $scCO_2$–mediated drug impregnation process a simple depressurization step allows a ready-to-use organic-solvent-free drug loaded material to be obtained. Furthermore, it offers the possibility to tailor the operating parameters of the impregnation process, such as temperature, pressure, and time, on the basis of the selected drug/carrier system [22]. This permits a better drug/carrier interaction to be obtained, with the drug in an amorphous state, which improves its dissolution profile and, consequently, its bioavailability [23].

Notwithstanding the above reported remarkable advantages, some drawbacks in the use of this technology have emerged, such as the scarce ability of $scCO_2$ to dissolve polar and ionic species, since it is a linear molecule with no net dipole moment. Furthermore, the elevated pressure required and high maintenance cost can represent a limitation in the use of this technology [21].

Even though the incorporation of active pharmaceutical ingredients (APIs) in organic and inorganic carriers through $scCO_2$ has been proposed in different research areas [22,24,25], the loading of drugs on ZnO carriers has not been investigated yet [23] and, to the best of our knowledge, the clotrimazole incorporation described in our previous work [26] represents the very first study about the loading of an API on nanostructured ZnO by means of $scCO_2$.

The fundamental idea of this research project is to combine the physico–chemical and biological properties of ZnO with the eco-friendly features of the $scCO_2$–mediated drug impregnation process to develop a green multifunctional device for treating painful wounds. This consists of the combination of antibacterial and anti-inflammatory/analgesic action in a single delivery system. ZnO is particularly suitable for this role, because its nanostructure can be tailored to host drug molecules and because it can offer intrinsic antimicrobial activity [15]. Ibuprofen (IBU) has been selected as the drug to be hosted in the ZnO nanostructures, since it is one of the most commonly used and most frequently prescribed non-steroidal anti-inflammatory drug (NSAID) for oral and topical administration due to its prominent analgesic role [27,28], and it has already been employed to prepare innovative pain-reducing wound dressings [18,19]. Furthermore IBU has also been widely used in many scCO$_2$–mediated drug impregnation processes [23].

This work is a feasibility study aiming at investigating the possibility of loading IBU on different ZnO carriers by means of scCO$_2$ and checking the antimicrobial activity as well as the capability of the obtained system to release Zn^{2+} ions and the drug, which are essential requirements for the development of a multifunctional device for wound healing applications. Two nanostructured ZnO (NsZnO) powders with different morphologies and physico–chemical parameters were synthetized through wet organic-solvent-free processes [26] and characterized by means of powder X-ray diffraction, FESEM images, and nitrogen adsorption analysis. The samples were also characterized from a biological point of view; particularly, their antimicrobial activity against different microbial strains and the in vitro Zn^{2+} release profiles from the NsZnO matrices were evaluated. IBU was loaded on the NsZnO carriers with a scCO$_2$–mediated drug impregnation process and in vitro dissolution studies of the loaded drug were performed.

2. Materials and Methods

2.1. Materials

All the reagents for chemical synthesis, as well as ibuprofen, were purchased from Sigma-Aldrich and used as received without further purification. Carbon dioxide with a purity of 99.998% was supplied by SIAD (Italy). Bidistilled water was used throughout this study.

2.2. Synthesis of Nanostructured ZnO (NsZnO)

Two different NsZnO materials were synthesized as previously described [26], using two different organic solvent free processes: the first one was based on the use of sole water as the solvent, i.e., a chemical bath deposition (NsZnO-1) method, while the second was a soft-template sol–gel synthesis method (NsZnO-2).

2.3. Drug Adsorption from Supercritical Carbon Dioxide

The scCO$_2$–mediated drug loading was carried out using a procedure that had been developed in previous works [26,29]. It consisted of contacting the drug and each of the two NsZnO materials in a static atmosphere of scCO$_2$ at constant temperature and pressure (Figure 1). First, a pellet of the drug (100 mg) and a pellet of the NsZnO (100 mg) were prepared and introduced into a glass cylinder of 1 cm diameter. A disc of filter paper was placed between the two pellets to prevent their contact and guarantee an efficient recovery of the samples at the end of the drug loading process. The glass cylinder was placed inside a stainless-steel vessel, which was put in an oven that kept the entire system at constant temperature. Liquid CO$_2$ was used to fill the vessel, then the temperature was increased to 35 °C and additional CO$_2$ was pumped to reach the target pressure (10 MPa). The pump was coupled with a cryogenic bath to prevent cavitation. The system was maintained at the above-reported conditions for 12 h. At the end of the drug loading process, the on–off valve was opened, and the apparatus was brought back to atmospheric pressure by means of a heated restrictor valve.

The IBU-containing materials are denoted hereafter as IBU@NsZnO-1 and IBU@NsZnO-2.

Moreover, the two carriers as such were treated in the same conditions in the absence of IBU in the glass cylinder, in order to investigate the effect of the scCO$_2$ treatment on the NsZnO samples.

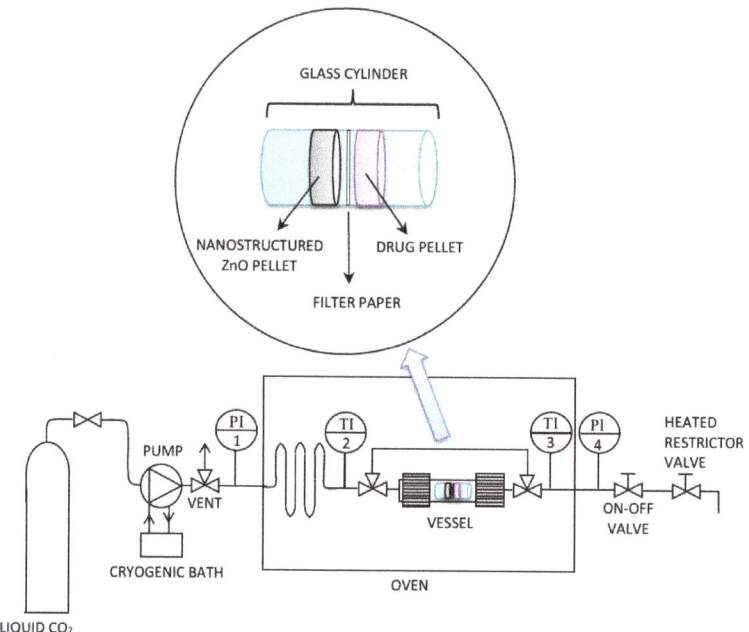

Figure 1. Experimental apparatus for supercritical carbon dioxide (scCO$_2$)-mediated drug loading.

2.4. Morphological Characterization

FESEM images were recorded with an FESEM Zeiss Merlin instrument, equipped with an EDS detector (Oxford Instruments, Abingdon-on-Thames, UK).

2.5. Powder X-ray Diffractometry

XRD patterns were obtained using a Panalytical X'Pert (Cu Kα radiation, Almelo, The Netherlands) diffractometer. Data were collected with a 2D solid state detector (PIXcel) from 20 to 70 2θ with a step size of 0.001 2θ and a wavelength of 1.54187 Å.

2.6. Nitrogen Adsorption Analysis

Nitrogen adsorption isotherms were measured using a Quantachrome AUTOSORB-1 instrument (Boynton Beach, FL, USA). Before the adsorption measurements, samples were outgassed for 2 h at 100 °C. BET specific surface areas (SSA$_{BET}$) were calculated in the relative pressure range of 0.04–0.1.

2.7. Thermogravimetry Analysis

Thermogravimetry (TG) analyses were carried out between 20 °C and 800 °C in air (flow rate 100 mL/min with a heating rate of 10 °C/min) using a SETARAM 92 instrument (Caluire et Cuire, France).

2.8. Antimicrobial Activity of NsZnO

2.8.1. Microbial Strains and Culture Conditions

The antibacterial activity of NsZnO-1 and NsZnO-2 was tested against a Gram-positive and a Gram-negative bacterial strain, such as *Staphylococcus aureus* ATCC 29213 and *Klebsiella pneumoniae* ATCC 700603, respectively. The antifungal activity of NsZnO-1 and NsZnO-2 samples was investigated against *Candida albicans* ATCC 90023. The strains were purchased from American Type Culture Collection (ATCC) (Manassas, VA, USA).

2.8.2. Inocula Preparation

Microorganism inocula were prepared by picking two to three colonies from an overnight culture of *S. aureus/K. pneumoniae* on Brain heart infusion agar (BHA, Merck KGaA, Darmstadt, Germany) or of *C. albicans* on Sabouraud dextrose (SAB, Merck KGaA, Darmstadt, Germany) agar at 37 °C (bacteria) or 35 °C (yeasts), suspending them in 5 mL of 0.85% normal saline, to yield a 0.5 McFarland turbidity standard, corresponding to a suspension of ~5 × 10^8 CFU/mL for bacteria or 5 × 10^6 CFU/mL for yeasts.

Bacterial suspensions were diluted 1:1000 in Mueller Hinton broth (MHB, Merck KGaA, Darmstadt, Germany) to obtain a final concentration of 10^5 CFU/mL. Fungal suspension was diluted 1:1000 in RPMI-1640 without sodium bicarbonate and with L-glutamine (Invitrogen, San Giuliano Milanese, Milano, Italy), buffered to pH 7.0 with 0.165 M morpholinepropanesulfonic acid (MOPS) (Sigma-Aldrich, Milan, Italy), and supplemented with glucose 0.2%, to obtain a concentration of 10^3 CFU/mL. Inocula were confirmed by colony counts in duplicate.

2.8.3. In Vitro Antimicrobial Assays

Determination of Minimum inhibitory concentration (MIC), Minimum bactericidal concentration (MBC), and Minimum fungicidal concentration (MFC).

The antimicrobial activity of NsZnO-1 and NsZnO-2 was determined using a broth microdilution (BM) method susceptibility assay, according to Clinical and Laboratory Standard Institute guidelines (CLSI document M07-A9 for bacteria, and M27-A3 for yeasts) [30,31]. As guidelines were not available for susceptibility testing of NsZnO, the antimicrobial activity was assessed following the CLSI BM method, with some modifications.

MIC determination was performed by serial dilution using 96-well microtiter plates (Sarstedt, Milan, Italy). Stock suspensions of NsZnO prepared at 30,000 µg/mL (w/v) in phosphate buffered solution (PBS; pH 7.4) were dispersed for 1 h using an ultrasonic bath, in order to minimize sedimentation of NsZnO particles. Doubling dilutions of the ZnO ranging from 15,000 to 30 µg/mL were prepared in 96-well microtiter plates in MHB for bacteria or in RPMI-1640 with MOPS for yeasts. After inoculum addition (0.1 mL), the plates were incubated under normal atmospheric conditions at 37 °C (bacteria) or 35 °C (yeasts) for 24 h. A sterile medium incubated under the same conditions was used as a blank, while the medium inoculated with the target microorganisms (without NsZnO) was used as a positive control of growth. All determinations were performed in duplicate. The lowest concentration of the NsZnO showing complete inhibition of visible growth was defined as MIC.

MBC and MFC of NsZnO were determined by subculturing 10 µL of broth taken from all the wells without visible growth onto BHA (bacteria) or SAB (yeasts) agar plates that did not contain the test agents. After incubation for 24 h at 37 °C (bacteria) or 35 °C (yeasts), MBC or MFC were defined as the lowest concentration of ZnO resulting in the death of 99.9% of the inoculum in no subculture [32].

Viable microorganism counts

To assess the antimicrobial activity of NsZO over time, the number of viable microorganisms was measured by monitoring bacterial/fungal growth after 24 h [33].

Briefly, the bacterial or yeast cells were grown overnight in BHI (Merck KGaA) or Sabouraud Dextrose (SAB, Merck KGaA, Darmstadt, Germany) broth at 37 °C or 35 °C, respectively. Bacteria and/or yeasts were harvested by centrifugation, washed, suspended in PBS, and diluted to yield a stock suspension of ~5 × 10^5 CFU/mL. All the NsZnO samples with concentration of 15,000 µg/mL, suspended in PBS, were incubated with bacterial or yeast suspension in a shaker incubator at 37 °C or 35 °C, respectively, for 24 h. PBS solution was used as a negative control. All samples were serially diluted and 100 µL of bacterial/yeasts suspension was drawn from each sample tube, spread on BHA or SAB agar plates, and incubated at 37 °C or 35 °C for 24 h, so that the number of CFU/mL could be determined.

2.9. In Vitro Zinc Ions Release

Zinc ion release from the samples NsZnO-1 and NsZnO-2 was studied using vertical Franz diffusion cells and synthetic skin (Dow Corning, 7-4107, Silicone Elastomer Membrane, Biesterfeld Polychem, Milan, Italy). Suspensions of NsZnO (5 mg of powder in 0.5 mL of PBS buffer solution) were employed as the donor phases. The receiving phase consisted of a PBS buffer solution at pH 7.4. The apparatus was maintained under stirring at 33 °C, during which, at scheduled time intervals, the receiving phase was withdrawn and entirely substituted with a fresh receiving phase. Zinc ion quantification was performed in each withdrawn sample using inductively coupled plasma mass spectrometry (ICP-MS, Thermo Scientific, Waltham, MA, USA).

2.10. Preliminary In Vitro Drug Release Study

The ability of the IBU-loaded NsZnO to release IBU was tested by using a multicompartment rotating cell equipped with a hydrophilic dialysis membrane (Spectra/Por, Spectrum®, cut-off 12,000–1,4000 Da, Sigma-Aldrich, Milan, Italy). PBS solution (1 mL) was used as the receiving medium. At predetermined time intervals, the receiving phase was completely replaced by a fresh solution, and analyzed for IBU content at 263 nm, using a Beckman–Coulter DU 730 Spectrophotometer (Indianapolis, IN, USA).

3. Results and Discussion

Figure 2 shows the FESEM pictures of NsZnO-1 and NsZnO-2. As observed in the previous study [26], the two carriers were characterized by different morphologies.

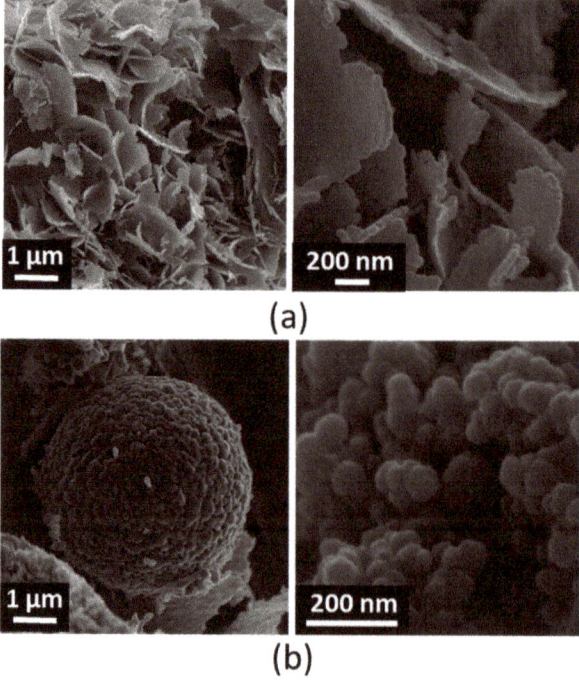

Figure 2. FESEM images of nanostructured (Ns)ZnO-1 (**a**) and NsZnO-2 (**b**).

NsZnO-1 appeared in the form of aggregates of nanosheets, with a thickness of about 20 nm, that were formed by the self-assembling of ovoid nanoparticles (having sizes around 15–20 nm). This morphology is in agreement with the mechanism growth proposed by Kakiuchi et al. [34].

The morphology of NsZnO-2 consisted of micrometric and sub-micrometric aggregates of nanoparticles with heterogeneous sizes of tens of nanometers.

Primary nanoparticles of NsZnO-1 were definitely smaller than those of NsZnO-2. The values of BET specific surface area and pore volume obtained by nitrogen adsorption are shown in Table 1. As previously observed [26], both features were larger for NsZnO-1 (68 m^2/g and 0.230 cm^3/g, respectively) than for NsZnO-2 (12 m^2/g and 0.050 cm^3/g, respectively), due to the lower particles size of NsZnO-1.

Table 1. Specific surface area (SSA) and pore volume values before and after the ibuprofen (IBU) adsorption by scCO$_2$ process.

	Before IBU Adsorption		After IBU Adsorption	
	SSA$_{BET}$ (m^2/g)	Pore Volume (cm^3/g)	SSA$_{BET}$ (m^2/g)	Pore Volume (cm^3/g)
NsZnO-1	68	0.230	8	0.04
NsZnO-2	12	0.050	nil	nil

Figure 3 reports the XRD patterns of both NsZnO materials, which reveal the occurrence of a highly crystalline single hexagonal phase of a wurtzite structure (JCPDS ICDD 36-1451). Comparing the XRD patterns of the two samples, it is evident that NsZnO-1 showed broader peaks than NsZnO-2, in agreement with the smaller particles size evidenced by FESEM analyses.

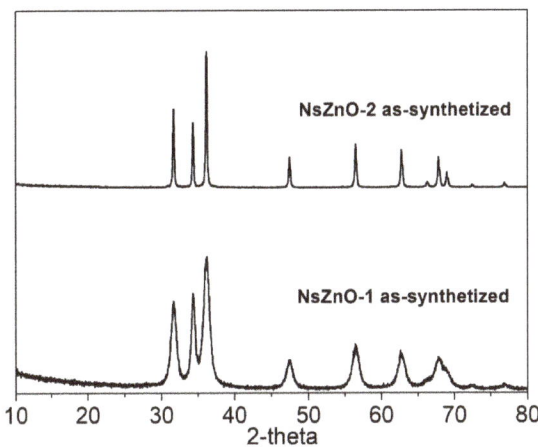

Figure 3. XRD patterns of NsZnO-1 and NsZnO-2.

In order to investigate the stability of NsZnO carriers in scCO$_2$, the XRD patterns of NsZnO-1 and NsZnO-2 after treatment in scCO$_2$ for 12 h, at 35 °C and 10 MPa, were collected and these are reported in Figure 4. In both cases the hexagonal wurtzite pattern of ZnO was preserved, and no new peaks were detected. This result showed that no extensive reaction between ZnO and the CO$_2$ occurred, which should not have been taken for granted, considering that the reaction between ZnO and CO$_2$ to give ZnCO$_3$ is a well-known phenomenon [35,36]. This evidence confirms the feasibility of using scCO$_2$ as a solvent for the drug loading of NsZnO carriers.

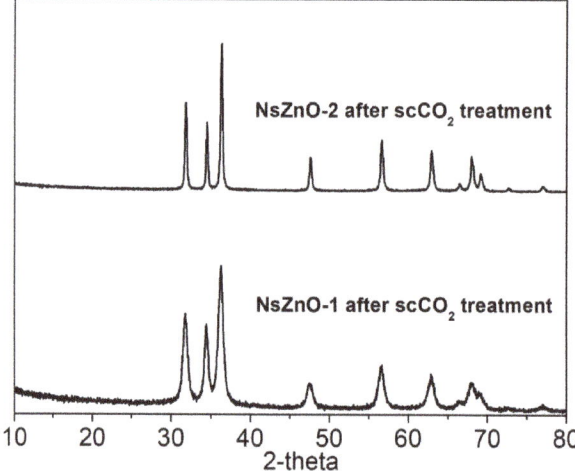

Figure 4. XRD patterns of NsZnO-1 and NsZnO-2 after scCO$_2$ treatment.

As far as the IBU loading is concerned, its content was calculated as the weight loss by TG analysis (Figure S1) and was found equal to 14% w/w for IBU@NsZnO-1 and 9% w/w for IBU@NsZnO-2, respectively (Table 2).

Table 2. Ibuprofen content in IBU@NsZnO-1 and IBU@NsZnO-2.

	IBU Content (% w/w)
IBU@NsZnO-1	14
IBU@NsZnO-2	9

The larger IBU content in IBU@NsZnO-1 than in IBU@NsZnO-2 was ascribed to the larger specific surface area and pore volume of NsZnO-1, which yielded larger drug adsorption and loading capacity.

Due to the presence of IBU molecules on the NsZnO carriers, the specific surface area and pore volume drastically decreased in both systems, as revealed by the data reported in Table 1.

It is worth noting that IBU contents in the two systems were similar to those previously obtained for clotrimazole adsorbed by scCO$_2$ on NsZnO-1 and NsZnO-2 carriers [26], which were equal to 17% w/w and 14% w/w, respectively. This suggests the robustness of the scCO$_2$ approach in the drug loading of NsZnOs.

XRD analyses were carried out to characterize IBU@NsZnO-1 and IBU@NsZnO-2. Figure 5 reports the XRD patterns of both systems, in comparison with those of the materials as-such and the pure crystalline IBU. No additional diffraction peaks typical of the crystalline IBU were observed for either IBU@NsZnO-1 or IBU@NsZnO-2 samples. This reveals that drug molecules are not assembled in the crystalline form on the two carriers.

The same result was previously obtained in the case of clotrimazole [26]. The amorphous form of the drug adsorbed on NsZnO from scCO$_2$ may be ascribed to the scCO$_2$–mediated process, which is known to favor the amorphization of the adsorbed drug [23]. This is a crucial aspect, in particular for poorly water-soluble drugs, because it is widely accepted that amorphization of the drug molecules plays a key role in increasing their dissolution rate and solubility.

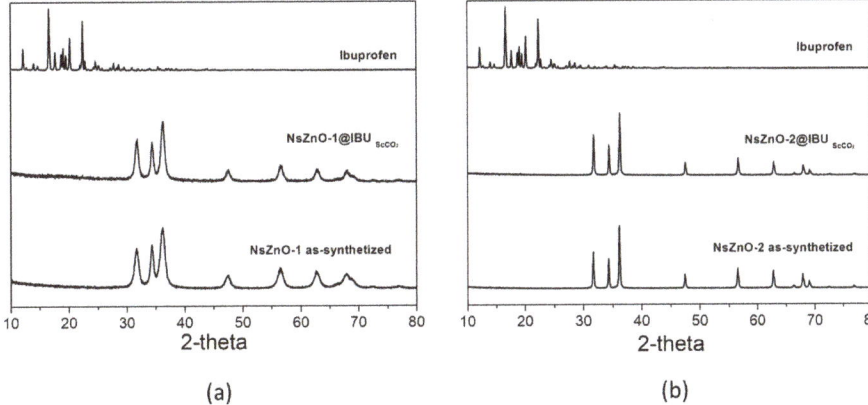

Figure 5. (a) XRD patterns of NsZnO-1, IBU@NsZnO-1, and pure crystalline IBU. (b) XRD patterns of NsZnO-2, IBU@NsZnO-2, and pure crystalline IBU.

In order to investigate the antimicrobial activity of NsZnOs, some microbiological parameters, such as MIC, MBC, and MFC were used (Table 3). In addition, a CFU assay was used to measure the antimicrobial activity of the NsZnOs over time by monitoring bacterial/fungal growth within 24 h (Table 4, and Figure 6).

Table 3. Minimum inhibitory concentration (MIC), and minimum bactericidal concentration (MBC), of NsZnO-1 and NsZnO-2 determined for *S. aureus*, *K. pneumoniae*, and *C. albicans* expressed in µg/mL. Minimum fungicidal concentration (MFC) for *C. albicans* was not determined.

Microbial Strain	S. aureus		K. pneumoniae		C. albicans
	MIC (µg/mL)	MBC (µg/mL)	MIC (µg/mL)	MBC (µg/mL)	MIC (µg/mL)
NsZnO-1	120	>470	470	1875	>15,000
NsZnO-2	230	>470	930	>3750	>15,000

As emerged from data shown in Table 3 the two ZnO nanostructures exhibited a stronger activity on the Gram-positive *S. aureus* than the Gram-negative *K. pneumoniae*. Between the two ZnO nanostructures, NsZnO-1 showed better activity than NsZnO-2 against *S. aureus* with an MIC value of 120 µg/mL vs 230 µg/mL. Since ZnO suspensions appeared "cloudy" in the case of *C. albicans*, it was not possible to determine the MIC from the visual appearance of the medium; hence, MFC was not assessed for this yeast (data not shown). In general, MBCs were two concentrations higher than MICs, with exception of NsZnO-2 that showed a MBC value one concentration higher than MIC against *S. aureus*, suggesting a more bacteriostatic activity of these compounds.

Table 4 and Figure 6 report the results of the viable microorganism counts assessed through a CFU assay. The bactericidal activity of NsZnO-1 against *S. aureus* (expressed in Log CFU/mL) was greater than that achieved by NsZnO-2 (1 vs 4.22, Figure 6a). The same trend, even if with less microbial load reduction values, is evident in Figure 6b for *K. pneumoniae*, where the Log CFU/mL of bacterial load was 7.21 and 8.42 for NsZnO-1 and NsZnO-2, respectively. Despite the failure of the broth dilution technique for yeasts, the enumeration of viable organisms' method was efficient in the determination of the antifungal activity. The results are shown in Figure 6c. NsZnO-1 log counts observed for *C. albicans* was 5.27, whereas NsZnO-2 was able to reduce the yeast cells growth of 6.15 log in comparison with ZnO-free controls (7.02 log). Taken together, these results indicate that the two ZnO nanostructures exhibited a better activity towards *S. aureus* than *K. pneumoniae* and *C. albicans*.

Table 4. Comparison of antibacterial/antifungal activity of NsZnO-1 and NsZnO-2 against *S. aureus*, *K. pneumoniae*, and *C. albicans* determined with the enumeration of viable microorganism assay and expressed in CFU/mL after 24 h of incubation.

Microbial Strain	*S. aureus*	*K. pneumoniae*	*C. albicans*
NsZnO-1	10 CFU/mL	1.62×10^7 CFU/mL	1.85×10^5 CFU/mL
NsZnO-2	1.65×10^4 CFU/mL	2.66×10^8 CFU/mL	1.43×10^6 CFU/mL

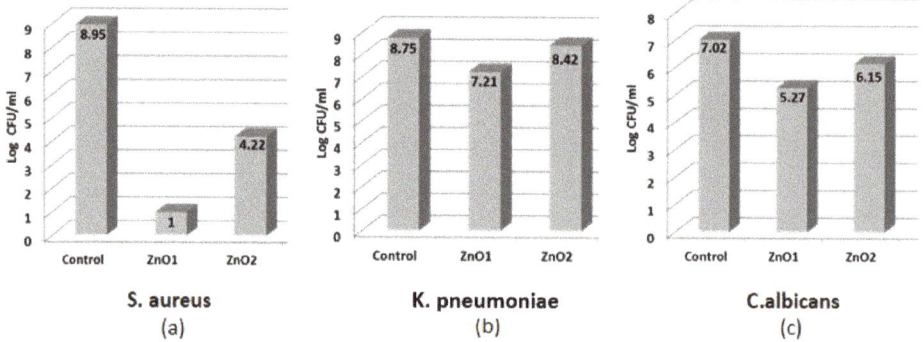

Figure 6. Comparison of antibacterial activities of NsZnO-1 and NsZnO-2 against *S. aureus* (**a**), *K. pneumoniae* (**b**), and *C. albicans* (**c**) expressed in Log CFU/mL.

Our data are difficult to compare due to the different methods and microorganisms used in the antimicrobial activity determination. However, these data are in agreement with those of some authors who detected a better antimicrobial activity of ZnO-compounds on Gram positive than Gram negative bacteria, and a good antifungal activity on *C. albicans* [13,37,38].

In addition, our results agree with the conclusions of Reddy et al. [39] and Tayel et al. [40] and disagree with the conclusions of Pasquet et al. [14]. In detail, Reddy and Tayel explained that the peptidoglycan cell-wall of Gram-positive bacteria may promote ZnO attachment onto the cell wall, whereas cell-wall lipophilic components of Gram-negative bacteria may oppose this attachment. Until now, the antifungal activity mechanism has not been well clarified; however, the candidacidal mechanism of ZnO can be probably ascribed to the cellular structure disruption or to inhibition of biological molecular synthesis due to Zn^{2+} release [38].

Among the NsZnOs investigated, the sample NsZnO-1 showed higher antimicrobial activity compared to NsZnO-2. This trend was confirmed by both the in vitro tests. This phenomenon could be ascribed to the crystallite sizes of the nanoparticles, which have been reported to greatly impact their antimicrobial activity, probably because of a greater accumulation of the nanoparticles inside the cell membrane and cytoplasm [12]. In fact, NsZnO-1 is characterized by lower crystallite sizes than NsZnO-2. This observation can be reinforced by the results obtained with *C. albicans*, against which NsZnO-1 showed better antifungal activity than NsZnO-2. These conclusions are consistent with the study of Lipovsky et al. [41], who suggested that ZnO nanoparticles display a marked activity against *C. albicans* and that the cytotoxic effect is size dependent.

Among the key mechanisms influencing the antimicrobial activity of nanostructured ZnO, it is important to consider the release of Zn^{2+} ions.

For this reason, a simple test was carried out to study the zinc ion release from NsZnO-1 and NsZnO-2 using vertical Franz diffusion cells equipped with synthetic skin. The results drawn from the zinc ion quantification are shown in Figure 7. The in vitro Zn^{2+} release study evidenced the ability of both the NsZnOs to release Zn^{2+}, highlighting their potential use as multifunctional antimicrobial drug carriers.

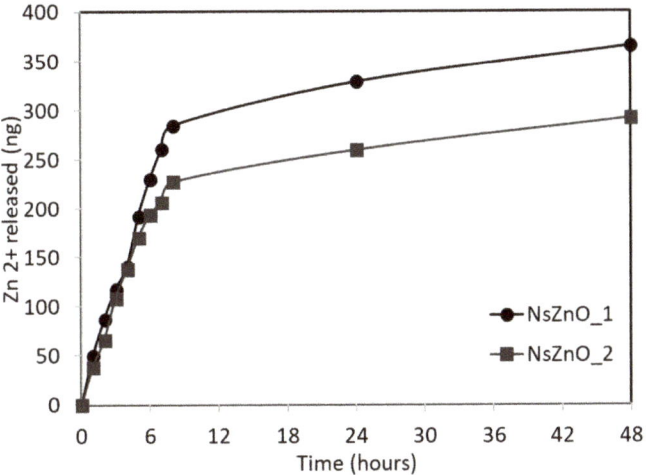

Figure 7. In vitro Zn^{2+} release from NsZnO-1 and NsZnO-2.

A higher amount of Zn^{2+} ion was released from NsZnO-1 and this was ascribed to the lower crystallite size and the higher SSA of the sample. The maximum amount of released Zn^{2+} ion after 48 h corresponded to a percentage by mass of zinc equal to about 0.009% for NsZnO-1 and 0.007% for NsZnO-2: these low values confirm that the release is a surface phenomenon.

The higher Zn^{2+} ion release ability of NsZnO-1 is in agreement with the higher antibacterial activity of this carrier (Figure 6).

The in vitro ibuprofen release study was aimed at verifying the possibility of releasing the drug from IBU@NsZnO-1 and IBU@NsZnO-2 systems, assessing the lack of complete irreversible trapping of drug molecules in the carrier. The same test was carried out with crystalline ibuprofen for comparison.

Figure 8 shows the cumulative release curves of ibuprofen. The percentage of drug released in 6 h was 68% for IBU@NsZnO-2, 44% for IBU@NsZnO-1, and 57% for crystalline ibuprofen.

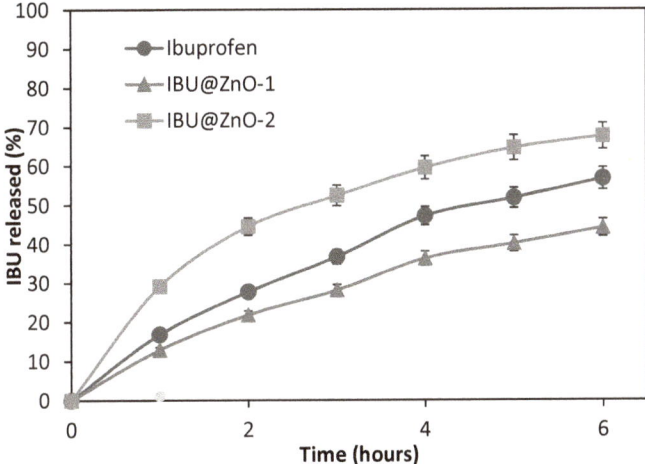

Figure 8. In vitro release profile of ibuprofen from IBU@NsZnO-1, IBU@NsZNO-2, and crystalline ibuprofen.

These data reveal that both NsZnO-1 and NsZnO-2 were able to act as carriers for ibuprofen delivery. The different percentage of IBU released from the two materials may be ascribed to the different morphology and pore volume, which affect the distribution of drug molecules in the carrier and their diffusion to the receiving solution.

In conclusion, this preliminary in vitro release test showed that ibuprofen adsorbed on the NsZnO-1 and NsZnO-2 by scCO$_2$ can be delivered, confirming the potential role of these nanostructures as drug delivery systems, as previously observed in the case of clotrimazole [26].

4. Conclusions

A feasibility study was conducted to investigate the possibility of developing a green multifunctional device for wound healing applications. Two multifunctional drug delivery systems based on nanostructured ZnO were prepared by means of non-toxic and organic-solvent free procedures.

The antimicrobial properties of the ZnO carriers were investigated against both Gram-positive and Gram-negative bacterial strains, such as *S. aureus* and *K. pneumoniae*, as well as against *C. albicans*. As a whole, the results indicated that the two ZnO nanostructures exhibited the following activity: *S. aureus* > *C. albicans* > *K. pneumoniae*.

Moreover, an in vitro Zn^{2+} release study was carried out. A correlation between the antimicrobial activity, the physico–chemical properties (specific surface area and crystal size) of nanostructured ZnO and the Zn^{2+} ion release was found.

For the first time ibuprofen was successfully loaded on the nanostructured ZnO carriers with a supercritical CO$_2$-mediated drug impregnation process. The drug-loaded amount was observed to depend on the specific surface area and the pore volume of the carrier and up to 14% *w/w* of ibuprofen in its amorphous form was obtained inside the final drug delivery system. A preliminary drug release test showed that up to 68% of the loaded ibuprofen could be delivered to a biological medium, confirming the feasibility of using nanostructured ZnO as a multifunctional antimicrobial drug carrier.

Supplementary Materials: The following are available online at http://www.mdpi.com/2079-4991/9/3/407/s1, Figure S1: Section A: TG curves of NsZnO-1 and IBU@NsZnO-1; Section B: TG curves of NsZnO-2 and IBU@NsZnO-2.

Author Contributions: Conceptualization, F.L. and B.O.; Investigation, F.L., R.C. (Roberta Cataldo), S.S.Y.M., S.R., and N.M.; Methodology, L.M., M.B., and N.M.; Resources, L.M., R.C. (Roberta Cavalli), and B.O.; Supervision, B.O.; Validation, L.M., S.R., V.T., and R.C. (Roberta Cavalli); Writing—original draft, F.L. and M.B.; Writing—review and editing, S.S.Y.M., M.B., V.T., and B.O.

Funding: This research received no external funding.

Conflicts of Interest: The authors declare no conflict of interest.

References

1. Kolodziejczak-Radzimska, A.; Jesionowski, T. Zinc oxide-from synthesis to application: A review. *Materials (Basel)* **2014**, *7*, 2833–2881. [CrossRef] [PubMed]
2. Wang, Z.L.; Kong, X.Y.; Ding, Y.; Gao, P.; Hughes, W.L.; Yang, R.; Zhang, Y. Semiconducting and piezoelectric oxide nanostructures induced by polar surfaces. *Adv. Funct. Mater.* **2004**, *14*, 943–956. [CrossRef]
3. Wang, Z.L. Novel nanostructures of ZnO for nanoscale photonics, optoelectronics, piezoelectricity, and sensing. *Appl. Phys. A Mater. Sci. Process.* **2007**, *88*, 7–15. [CrossRef]
4. Pasquet, J.; Chevalier, Y.; Couval, E.; Bouvier, D.; Bolzinger, M.A. Zinc oxide as a new antimicrobial preservative of topical products: Interactions with common formulation ingredients. *Int. J. Pharm.* **2015**, *479*, 88–95. [CrossRef] [PubMed]
5. Sirelkhatim, A.; Mahmud, S.; Seeni, A.; Kaus, N.H.M.; Ann, L.C.; Bakhori, S.K.M.; Hasan, H.; Mohamad, D. Review on zinc oxide nanoparticles: Antibacterial activity and toxicity mechanism. *Nano-Micro Lett.* **2015**, *7*, 219–242. [CrossRef] [PubMed]

6. Mirzaei, H.; Darroudi, M. Zinc oxide nanoparticles: Biological synthesis and biomedical applications. *Ceram. Int.* **2017**, *43*, 907–914. [CrossRef]
7. Zafar, R.; Zia, K.M.; Tabasum, S.; Jabeen, F.; Noreen, A.; Zuber, M. Polysaccharide based bionanocomposites, properties and applications: A review. *Int. J. Biol. Macromol.* **2016**, *92*, 1012–1024. [CrossRef] [PubMed]
8. Díez-Pascual, A.M.; Díez-Vicente, A.L. Wound healing bionanocomposites based on castor oil polymeric films reinforced with chitosan-modified ZnO nanoparticles. *Biomacromolecules* **2015**, *16*, 2631–2644. [CrossRef] [PubMed]
9. Yadollahi, M.; Gholamali, I.; Namazi, H.; Aghazadeh, M. Synthesis and characterization of antibacterial carboxymethyl Chitosan/ZnO nanocomposite hydrogels. *Int. J. Biol. Macromol.* **2015**, *74*, 136–141. [CrossRef] [PubMed]
10. Staneva, D.; Atanasova, D.; Vasileva-Tonkova, E.; Lukanova, V.; Grabchev, I. A cotton fabric modified with a hydrogel containing ZnO nanoparticles. Preparation and properties study. *Appl. Surf. Sci.* **2015**, *345*, 72–80. [CrossRef]
11. Raguvaran, R.; Manuja, B.K.; Chopra, M.; Thakur, R.; Anand, T.; Kalia, A.; Manuja, A. Sodium alginate and gum acacia hydrogels of ZnO nanoparticles show wound healing effect on fibroblast cells. *Int. J. Biol. Macromol.* **2017**, *96*, 185–191. [CrossRef] [PubMed]
12. Jones, N.; Ray, B.; Ranjit, K.T.; Manna, A.C. Antibacterial activity of ZnO nanoparticle suspensions on a broad spectrum of microorganisms. *FEMS Microbiol. Lett.* **2008**, *279*, 71–76. [CrossRef] [PubMed]
13. Khezerlou, A.; Alizadeh-sani, M.; Azizi-lalabadi, M.; Ehsani, A. Microbial Pathogenesis Nanoparticles and their antimicrobial properties against pathogens including bacteria, fungi, parasites and viruses. *Microb. Pthogenes.* **2018**, *123*, 505–526. [CrossRef] [PubMed]
14. Pasquet, J.; Chevalier, Y.; Couval, E.; Bouvier, D.; Noizet, G.; Morlière, C.; Bolzinger, M.A. Antimicrobial activity of zinc oxide particles on five micro-organisms of the Challenge Tests related to their physicochemical properties. *Int. J. Pharm.* **2014**, *460*, 92–100. [CrossRef] [PubMed]
15. Kumar, R.; Umar, A.; Kumar, G.; Nalwa, H.S. Antimicrobial properties of ZnO nanomaterials: A review. *Ceram. Int.* **2017**, *43*, 3940–3961. [CrossRef]
16. Agren, M.S.; Mirastschijski, U. The release of zinc ions from and cytocompatibility of two zinc oxide dressings. *J. Wound Care* **2004**, *13*, 367–369. [PubMed]
17. Lansdown, A.B.G.; Mirastschijski, U.; Stubbs, N.; Scanlon, E.; Ågren, M.S. Zinc in wound healing: Theoretical, experimental, and clinical aspects. *Wound Repair Regen.* **2007**, *15*, 2–16. [CrossRef] [PubMed]
18. Morgado, P.I.; Miguel, S.P.; Correia, I.J.; Aguiar-Ricardo, A. Ibuprofen loaded PVA/chitosan membranes: A highly efficient strategy towards an improved skin wound healing. *Carbohydr. Polym.* **2017**, *159*, 136–145. [CrossRef] [PubMed]
19. Price, P.; Fogh, K.; Glynn, C.; Krasner, D.L.; Osterbrink, J.; Sibbald, R.G. Why combine a foam dressing with ibuprofen for wound pain and moist wound healing? *Int. Wound J.* **2007**, *4*, 1–3. [CrossRef] [PubMed]
20. Xiong, H.M. ZnO nanoparticles applied to bioimaging and drug delivery. *Adv. Mater.* **2013**, *25*, 5329–5335. [CrossRef] [PubMed]
21. Girotra, P.; Singh, S.K.; Nagpal, K. Supercritical fluid technology: A promising approach in pharmaceutical research. *Pharm. Dev. Technol.* **2012**, *18*, 22–38. [CrossRef] [PubMed]
22. Champeau, M.; Thomassin, J.M.; Tassaing, T.; Jérôme, C. Drug loading of polymer implants by supercritical CO_2 assisted impregnation: A review. *J. Control. Release* **2015**, *209*, 248–259. [CrossRef] [PubMed]
23. Gurikov, P.; Smirnova, I. Amorphization of drugs by adsorptive precipitation from supercritical solutions: A review. *J. Supercrit. Fluids* **2018**, *132*, 105–125. [CrossRef]
24. Banchero, M.; Ronchetti, S.; Manna, L. Characterization of ketoprofen/methyl-β-cyclodextrin complexes prepared using supercritical carbon dioxide. *J. Chem.* **2013**, *2013*, 583952. [CrossRef]
25. Rudrangi, S.R.S.; Trivedi, V.; Mitchell, J.C.; Wicks, S.R.; Alexander, B.D. Preparation of olanzapine and methyl-β-cyclodextrin complexes using a single-step, organic solvent-free supercritical fluid process: An approach to enhance the solubility and dissolution properties. *Int. J. Pharm.* **2015**, *494*, 408–416. [CrossRef] [PubMed]
26. Leone, F.; Gignone, A.; Ronchetti, S.; Cavalli, R.; Manna, L.; Banchero, M.; Onida, B. A green organic-solvent-free route to prepare nanostructured zinc oxide carriers of clotrimazole for pharmaceutical applications. *J. Clean. Prod.* **2018**, *172*, 1433–1439. [CrossRef]

27. Bushra, R.; Aslam, N. An overview of clinical pharmacology of Ibuprofen. *Oman Med. J.* **2010**, *25*, 155–161. [PubMed]
28. Irvine, J.; Afrose, A.; Islam, N. Formulation and delivery strategies of ibuprofen: challenges and opportunities. *Drug Dev. Ind. Pharm.* **2018**, *44*, 173–183. [CrossRef]
29. Gignone, A.; Manna, L.; Ronchetti, S.; Banchero, M.; Onida, B. Microporous and Mesoporous Materials Incorporation of clotrimazole in Ordered Mesoporous Silica by supercritical CO_2. *Microporous Mesoporous Mater.* **2014**, *200*, 291–296. [CrossRef]
30. *Reference Method for Broth Dilution Antifungal Susceptibility Testing of Yeasts*, Approved Standard, 3rd ed.; Clinical and Laboratory Standards Institute: Wayne, PA, USA, 2008; Volume 28.
31. *Methods for Dilution Antimicrobial Susceptibility Tests for Bacteria That Grow Aerobically*, Approved Standard, 9th ed.; Clinical and Laboratory Standards Institute: Wayne, PA, USA, 2012; Volume 32.
32. Mandras, N.; Nostro, A.; Roana, J.; Scalas, D.; Banche, G.; Ghisetti, V.; Del Re, S.; Fucale, G.; Cuffini, A.M.; Tullio, V. Liquid and vapour-phase antifungal activities of essential oils against Candida albicans and non-albicans Candida. *BMC Complement. Altern. Med.* **2016**, *16*, 1–7. [CrossRef] [PubMed]
33. Wahid, F.; Yin, J.J.; Xue, D.D.; Xue, H.; Lu, Y.S.; Zhong, C.; Chu, L.Q. Synthesis and characterization of antibacterial carboxymethyl Chitosan/ZnO nanocomposite hydrogels. *Int. J. Biol. Macromol.* **2016**, *88*, 273–279. [CrossRef] [PubMed]
34. Kakiuchi, K.; Hosono, E.; Kimura, T.; Imai, H.; Fujihara, S. Fabrication of mesoporous ZnO nanosheets from precursor templates grown in aqueous solutions. *J. Sol-Gel Sci. Technol.* **2006**, *39*, 63–72. [CrossRef]
35. Anas, M.; Gönel, A.G.; Bozbag, S.E.; Erkey, C. Thermodynamics of Adsorption of Carbon Dioxide on Various Aerogels. *J. CO2 Util.* **2017**, *21*, 82–88. [CrossRef]
36. Graedel, T.E. Corrosion Mechanisms for Zinc Exposed to the Atmosphere. *J. Electrochem. Soc.* **1989**, *136*, 193C–203C. [CrossRef]
37. Ramani, M.; Ponnusamy, S.; Muthamizhchelvan, C.; Marsili, E. Amino acid-mediated synthesis of zinc oxide nanostructures and evaluation of their facet-dependent antimicrobial activity. *Colloids Surf. B Biointerfaces* **2014**, *117*, 233–239. [CrossRef] [PubMed]
38. Sun, Q.; Li, J.; Le, T. Zinc Oxide Nanoparticle as a Novel Class of Antifungal Agents: Current Advances and Future Perspectives. *J. Agric. Food Chem.* **2018**, *66*, 11209–11220. [CrossRef] [PubMed]
39. Reddy, K.M.; Feris, K.; Bell, J.; Hanley, C.; Punnoose, A. Selective toxicity of zinc oxide nanoparticles to prokaryotic and eukaryotic systems. *Appl. Phys. Lett.* **2007**, *24*, 213902-1–213902-3. [CrossRef] [PubMed]
40. Tayel, A.A.; El-Tras, W.F.; Moussa, S.; El-Baz, A.F.; Mahrous, H.; Salem, M.F.; Brimer, L. Antibacterial action of zinc oxide nanoparticles against foodborne pathogens. *J. Food Saf.* **2011**, *31*, 211–218. [CrossRef]
41. Lipovsky, A.; Nitzan, Y.; Gedanken, A.; Lubart, R. Antifungal activity of ZnO nanoparticles-the role of ROS mediated cell injury. *Nanotechnology* **2011**, *22*, 105101. [CrossRef] [PubMed]

© 2019 by the authors. Licensee MDPI, Basel, Switzerland. This article is an open access article distributed under the terms and conditions of the Creative Commons Attribution (CC BY) license (http://creativecommons.org/licenses/by/4.0/).

MDPI
St. Alban-Anlage 66
4052 Basel
Switzerland
Tel. +41 61 683 77 34
Fax +41 61 302 89 18
www.mdpi.com

Nanomaterials Editorial Office
E-mail: nanomaterials@mdpi.com
www.mdpi.com/journal/nanomaterials

www.ingramcontent.com/pod-product-compliance
Lightning Source LLC
LaVergne TN
LVHW070546100526
838202LV00012B/395